CONTENTS

Targeting Maths Homework Year 3

ISBN: 978 1 925726 56 5

Published by Pascal Press
PO Box 250
Glebe NSW 2037
www.pascalpress.com.au
contact@pascalpress.com.au

Author: Ann & Johnny Baker
Publisher: Lynn Dickinson
Editor: Ruth Schultz
Cover and Text Designer: Janice Bowles

Typesetter: Ruth Schultz
Images & Illustrations: Dreamstime (unless otherwise indicated)

Printed by Wai Man Bookbinding (China) Ltd

Introduction

Targeting Maths Homework aims to build and reinforce Maths skills. This book supports the ACARA Australian Version 9·0 Curriculum for Year 3 and helps children to revise and consolidate what has been taught in the classroom. ACARA codes are shown on each unit and a chart explaining their content descriptions is on pages v and vi. The inside cover shows the topics in each unit.

The structure of this book

This book has 32 carefully graded four-page units.
The units are divided between:

- ★ Number and Algebra
- ★ Measurement and Space
- ★ Statistics & Probability
- ★ Problem Solving.

Assessment

Term Reviews follow Units 1–8, 9–16, 17–24 and 25–32 to test work covered during the term, and allow parents and carers to monitor their child's progress. Children are encouraged to mark each unit as it is completed and to colour in the traffic lights at the end of each segment. These results are then transferred to the Marking Grid. Parents and carers can see at a glance if their child is excelling or struggling!

NAPLAN*-style questions

Each unit has a set of three NAPLAN*-style questions that directly match the learning in the unit and mirror question styles that frequently occur in NAPLAN. This is to ensure that children's familiarity and confidence with NAPLAN*-style questions is consistently developed.

● **Green** = Excellent — 2 or fewer questions incorrect
● **Orange** = Passing — 50% or more questions answered correctly
● **Red** = Struggling — fewer than 50% correct and needs help

SCORE	/10	0-4	5-8	9-10

How to Use This Book

The activities in this book are specifically designed to be used at home with minimal resources and support. Helpful explanations of key concepts and skills are provided throughout the book to help understand the tasks. Useful examples of how to do the activities are provided.

Regular practice of key concepts and skills will support the work your child does in school and will enable you to monitor their progress throughout the year. It is recommended that children complete 8 units per school term (one a week) and then the Term Review. Every unit has a Traffic Light scoreboard at the end of each section.

SCORE | /10 | 0-4 | 5-8 | 9-10

You or your child should mark each completed unit and then colour the traffic light that corresponds to the number of correct questions. This process will enable you to see at a glance how your child is progressing and to identify weak spots. The results should be recorded at the end of each term on the Marking Grid on page 1.

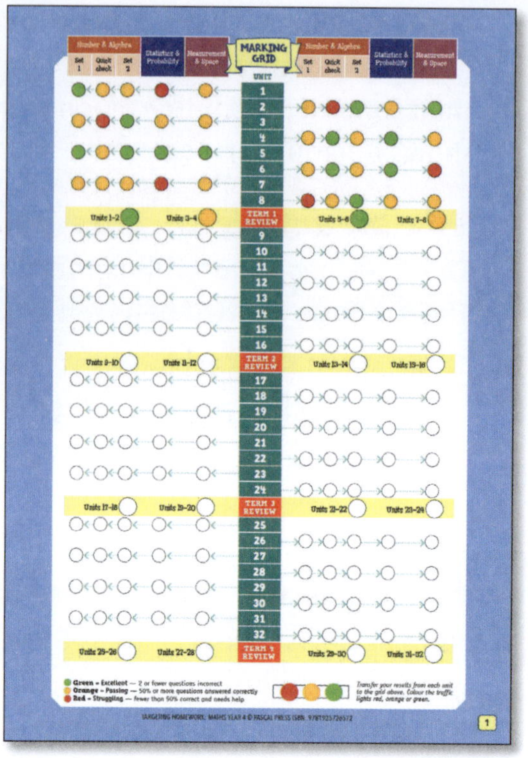

NOTE: The Problem of the Week sections do not appear on the Marking Grid as they often have multiple or subjective answers that cannot be easily scored.

The Term Review results are important for tracking progress and identifying any improvements in performance. If you find that certain questions are repeatedly causing difficulties and errors, then there is a good reason to discuss this with your child's teacher and arrange for extra instruction in that problem area.

Quick Maths Facts

The Quick Maths Facts on page 146 include multiplication tables up to 12 times, measurement equivalents, simple decimal–fraction–percentage conversions, a visual reminder of place value into the millions and the 'Days of the Month' rhyme. An accurate ruler has also been provided. Copy and laminate this page so that it is always available as a handy reference for your child.

Answers

The answer section on pages 147–161 can be removed, stapled together and kept somewhere safe. Use it to check answers when your child has completed each unit. Encourage your child to colour in the Traffic Light boxes when the answers have been calculated.

Australian Curriculum Correlations V9

Year 3 Maths		Number & Algebra	Statistics & Probability	Measurement & Space
ACARA CODE	**CONTENT DESCRIPTION**	**UNITS**	**UNITS**	**UNITS**
NUMBER				
AC9M3N01	Recognise, represent and order natural numbers using naming and writing conventions for numerals beyond 10 000	3, 4, 5, 11, 12, 13, 20, 25, 31		
AC9M3N02	Recognise and represent unit fractions including $\frac{1}{2}$, $\frac{1}{3}$, $\frac{1}{4}$, $\frac{1}{5}$ and $\frac{1}{10}$ and their multiples in different ways; combine fractions with the same denominator to complete the whole	2, 8, 10, 13, 18, 24, 26, 28		
AC9M3N03	Add and subtract two- and three-digit numbers using place value to partition, rearrange and regroup numbers to assist in calculations without a calculator	5, 6, 11, 12, 20, 28		
AC9M3N04	Multiply and divide one- and two-digit numbers, representing problems using number sentences, diagrams and arrays, and using a variety of calculation strategies	8, 14, 15, 23, 24, 29, 32		
AC9M3N05	Estimate the quantity of objects in collections and make estimates when solving problems to determine the reasonableness of calculations	9, 16, 21, 26, 27, 32		
AC9M3N06	Use mathematical modelling to solve practical problems involving additive and multiplicative situations including financial contexts; formulate problems using number sentences and choose calculation strategies, using digital tools where appropriate; interpret and communicate solutions in terms of the situation	4, 10, 15, 16, 26, 27, 31	32	
AC9M3N07	Follow and create algorithms involving a sequence of steps and decisions to investigate numbers; describe any emerging patterns	1, 6, 17, 25		
ALGEBRA				
AC9M3A01	Recognise and explain the connection between addition and subtraction as inverse operations, apply to partition numbers and find unknown values in number sentences	2, 17, 18, 22		
AC9M3A02	Extend and apply knowledge of addition and subtraction facts to 20 to develop efficient mental strategies for computation with larger numbers without a calculator	1, 2, 3, 21, 30		
AC9M3A03	Recall and demonstrate proficiency with multiplication facts for 3, 4, 5 and 10; extend and apply facts to develop the related division facts	3, 6, 7, 14, 15, 19, 22, 29, 30		

Australian CURRICULUM

Year 3 Maths		Number & Algebra	Statistics & Probability	Measurement & Space
ACARA CODE	**CONTENT DESCRIPTION**	**UNITS**	**UNITS**	**UNITS**
MEASUREMENT				
AC9M3M01	Identify which metric units are used to measure everyday items; use measurements of familiar items and known units to make estimates			2, 5, 13, 22, 24, 29
AC9M3M02	Measure and compare objects using familiar metric units of length, mass and capacity, and instruments with labelled markings			4, 9, 13, 16, 17, 24, 26
AC9M3M03	Recognise and use the relationship between formal units of time including days, hours, minutes and seconds to estimate and compare the duration of events			8, 11, 15, 19, 31
AC9M3M04	Describe the relationship between the hours and minutes on analog and digital clocks, and read the time to the nearest minute			8, 11, 21, 25, 30
AC9M3M05	Identify angles as measures of turn and compare angles with right angles in everyday situations			6, 14, 30
AC9M3M06	Recognise the relationships between dollars and cents and represent money values in different ways	5, 10	17	18, 32
SPACE				
AC9M3SP01	Make, compare and classify objects, identifying key features and explaining why these features make them suited to their uses			3, 5, 7, 10, 20, 27, 28
AC9M3SP02	Interpret and create two-dimensional representations of familiar environments, locating key landmarks and objects relative to each other			1, 9, 12, 23
STATISTICS				
AC9M3ST01	Acquire data for categorical and discrete numerical variables to address a question of interest or purpose by observing, collecting and accessing data sets; record the data using appropriate methods including frequency tables and spreadsheets			3, 4, 11, 15, 17, 18, 21, 27
AC9M3ST02	Create and compare different graphical representations of data sets including using software where appropriate; interpret the data in terms of the context			2, 8, 22, 23, 26, 28, 31
AC9M3ST03	Conduct guided statistical investigations involving the collection, representation and interpretation of data for categorical and discrete numerical variables with respect to questions of interest			13, 14, 32
PROBABILITY				
AC9M3P01	Identify practical activities and everyday events involving chance; describe possible outcomes and events as 'likely' or 'unlikely' and identify some events as 'certain' or 'impossible' explaining reasoning			1, 5, 6, 7, 10, 20, 25
AC9M3P02	Conduct repeated chance experiments; identify and describe possible outcomes, record the results, recognise and discuss the variation			9, 12, 16, 19, 24, 29, 30

Australian CURRICULUM

TARGETING MATHS HOMEWORK: YEAR 3 © PASCAL PRESS ISBN: 9781925726565

MARKING GRID

Number & Algebra			Statistics & Probability	Measurement & Space	UNIT	Number & Algebra			Statistics & Probability	Measurement & Space
Set 1	Quick check	Set 2				Set 1	Quick check	Set 2		
					1					
					2					
					3					
					4					
					5					
					6					
					7					
					8					
Units 1–2			Units 3–4		TERM 1 REVIEW	Units 5–6			Units 7–8	
					9					
					10					
					11					
					12					
					13					
					14					
					15					
					16					
Units 9–10			Units 11–12		TERM 2 REVIEW	Units 13–14			Units 15–16	
					17					
					18					
					19					
					20					
					21					
					22					
					23					
					24					
Units 17–18			Units 19–20		TERM 3 REVIEW	Units 21–22			Units 23–24	
					25					
					26					
					27					
					28					
					29					
					30					
					31					
					32					
Units 25–26			Units 27–28		TERM 4 REVIEW	Units 29–30			Units 31–32	

- **Green** = Excellent — 2 or fewer questions incorrect
- **Orange** = Passing — 50% or more questions answered correctly
- **Red** = Struggling — fewer than 50% correct and needs help

Transfer your results from each unit to the grid above. Colour the traffic lights red, orange or green.

AC9M3N07

Number Patterns and Sequences

Continue these sequences. The first one is done for you.

1. 2, 4, 6, 8, 10, 12, 14, 16 Going up in 2s
2. 3, 6, 9, 12, ____, ____, ____, ____ Going ________ in ____
3. 18, 15, 12, ____, ____, ____, ____ Going ________ in ____
4. 18, 24, 30, 36, ____, ____, ____, ____ Going ________ in ____
5. 90, 95, ____, ____, 110, 115, ____, ____ Going ________ in ____
6. 490, 495, ____, ____, 510, ____, ____ Going ________ in ____
7. 309, 307, 305, 303, ____, ____, ____, ____ Going ________ in ____
8. 976, 978, ____, ____, 984, ____, ____ Going ________ in ____
9. 111, 107, 103, ____, ____, 91, ____, ____ Going ________ in ____

Continue these number sequences. Say what changes and what stays the same in each sequence. The first one is done for you.

10. 13, 23, 33, 43, 53, 63
    The ones stay the same
    The tens go up by 1 ten
11. 5, 25, 45, ____, ____, ____
    The ones ________________
    The tens ________________
12. 56, 46, 36, ____, ____, ____
    The ones ________________
    The tens ________________
13. 76, 65, 54, ____, ____, ____
    The ones ________________
    The tens ________________
14. 102, 92, 82, ____, ____, ____
    The ones ________________
    The tens ________________
15. 76, 85, 94, ____, ____, ____
    The ones ________________
    The tens ________________

Now it's your turn to make up a number sequence.
Write the sequence. Say what stays the same and what changes.

16. ____, ____, ____, ____, ____
    The ones ________________. The tens ________________.

SCORE /16 0-7 8-14 15-16

Quick Check

Write the number that is 1 more.

1. 99 ______
2. 109 ______
3. 190 ______
4. 990 ______
5. 999 ______

Write the number that is 1 less.

6. 101 ______
7. 190 ______
8. 300 ______
9. 1000 ______
10. 10 000 ______

SCORE /10 0-4 5-8 9-10

TARGETING MATHS HOMEWORK: YEAR 3 © PASCAL PRESS ISBN: 9781925726565

Number & Algebra

AC9M3A02

Addition: Bridging Through 10

Sometimes you need to find the sum of two numbers, like 18 + 7, by bridging through 10.

For 18 + 7, split the 7 into 2 and 5.

This is how to show the **bridging through 10** on a number line:

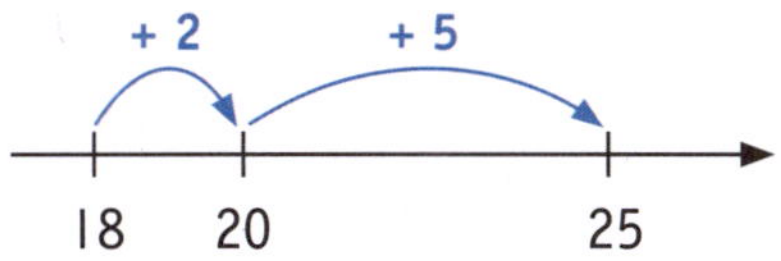

Show the bridging through 10 on the number line. Then write the answer.

① 17 + 6 = ____

③ 28 + 6 = ____

② 29 + 7 = ____

④ 37 + 8 = ____

Use bridging through 10 in your head for these additions.

⑤ 49 + 7 = ____

⑥ 78 + 5 = ____

⑦ 57 + 7 = ____

⑧ 69 + 7 = ____

⑨ 82 + 9 = ____

⑩ 49 + 6 = ____

⑪ 75 + 8 = ____

⑫ 36 + 9 = ____

⑬ 45 + 8 = ____

SCORE /13 0-6 7-11 12-13

Statistics & Probability

AC9M3P01

Probability in the Bag

Write the chances of each event happening. Choose from these probabilities: certain, very likely, quite likely, even chance, unlikely, impossible.

① Without looking, you take a red counter from the bag. ____________

② Without looking, you take a yellow counter from the bag. ____________

③ Without looking, you take a green counter from the bag. ____________

④ Without looking, you take a blue counter from the bag. ____________

⑤ How many red counters need to be added to the bag to make it an even chance of pulling out a red counter?

____

SCORE /5 0-2 3-4 5

AC9M3SP02

TERM 1

Position

Which letter is in each position?

A	B	C	D
E	F	G	H
I	J	K	L
M	N	O	P

1. To the right of I ☐
2. To the left of F ☐
3. Above O ☐
4. Between E and G ☐
5. 2 spaces to the left of D ☐
6. 3 spaces to the right of E ☐
7. 2 spaces down from C ☐
8. 3 spaces down from B ☐
9. In 4th place in the bottom row ☐
10. In 3rd place in the top row ☐

Write directions to move from A to K.

11. First move: ______________________

Second move: ______________________

SCORE /11 0-5 6-9 10-11

NAPLAN*-style Questions

1. There are 27 pieces of fruit in a basket. 19 are apples and the rest are oranges. How many oranges are there?

☐ oranges

2. The letter A was moved 3 spaces up, 4 spaces to the right and then 2 spaces down.

What letter is A covering now? ☐

P	Q	R	S	T
K	L	M	N	O
F	G	H	I	J
A	B	C	D	E

3. A box of chocolates has 3 strawberry, 2 toffee, 4 coffee and 5 mint chocolates in it.
Which flavour is the most likely to be picked, without looking?

◯ strawberry ◯ toffee
◯ coffee ◯ mint

SCORE /3 0-1 2 3

TARGETING MATHS HOMEWORK: YEAR 3 © PASCAL PRESS ISBN: 9781925726565

Problem of the Week

Patterns with Numbers

"Make 2 patterns with these numbers but do not use the same number twice," Mr Green told his class.

"What is the odd number out?" he asked.

11 9 12 5 23 2 17 25

3 21 15 8 6 14 20 18 24

Pattern 1

Pattern 2

The odd number out is:

"Too easy?" asked Mr Green.

"Try this one with 3 patterns. What is the odd number out in this set?

101 71 35 31 45 18 41 33

27 91 65 39 30 51 55 61

43 21 50 81 40 36 24 70 60

Pattern 3

Pattern 4

Pattern 5

The odd number out is:

1. Did you make your own set of numbers on pieces of paper? ◯ Yes ◯ No
2. If so, did this strategy help? ◯ Yes ◯ No
3. Did you use **Guess** and **Check**? ◯ Yes ◯ No
4. If so, did this strategy help? ◯ Yes ◯ No

Fractions

Count the parts to find out what fraction each circle shows.

1.  2 equal parts
halves

2.  ___ equal parts
____

3.  ___ equal parts
____

4.  ___ equal parts
____

5.  ___ equal parts
____

Name the fraction that is coloured blue. Then write it in numerical form.

6. 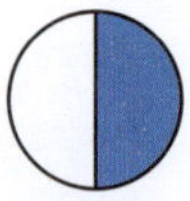 one-half is blue. $\frac{1}{2}$

7.  ___________ is blue. $\frac{\square}{\square}$

8. 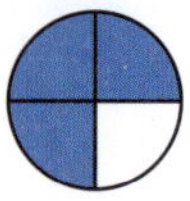 ___________ is blue. $\frac{\square}{\square}$

9.  ___________ is blue. $\frac{\square}{\square}$

10.  ___________ is blue. $\frac{\square}{\square}$

How many quarters are green?

11. 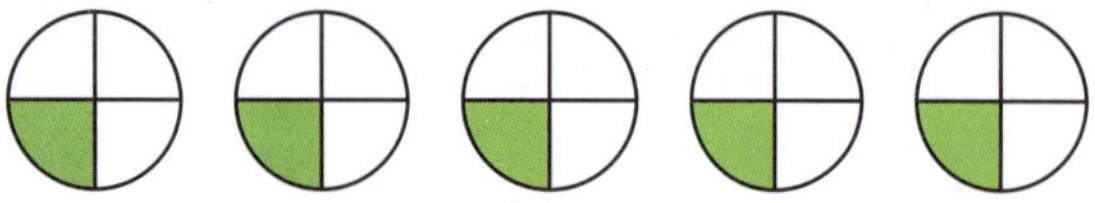 ☐

How many thirds are yellow?

12. 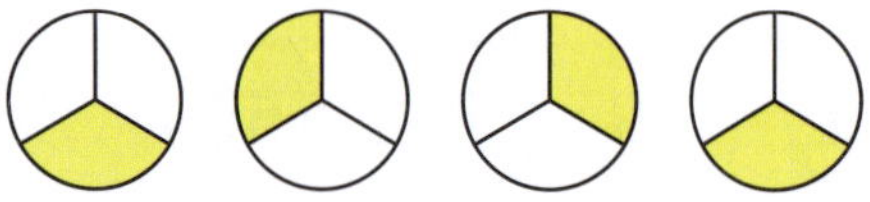 ☐

SCORE /12 0-5 6-10 11-12

Quick Check

Write these fractions in words.

1. $\frac{1}{4}$ ___________
2. $\frac{5}{8}$ ___________
3. $\frac{2}{3}$ ___________
4. $\frac{2}{5}$ ___________
5. $\frac{3}{4}$ ___________
6. $\frac{5}{6}$ ___________

Write these fractions as numbers.

7. one-third $\frac{\square}{\square}$
8. two-fifths $\frac{\square}{\square}$
9. three-quarters $\frac{\square}{\square}$
10. seven-eighths $\frac{\square}{\square}$

SCORE /10 0-4 5-8 9-10

TARGETING MATHS HOMEWORK: YEAR 3 © PASCAL PRESS ISBN: 9781925726565

Number & Algebra

AC9M3A01, AC9M3A02

Addition and Subtraction

You can use addition to help with subtraction.
If you know that **8 + 4 is 12**, then you also know that **12 – 8 = 4** and **12 – 4 = 8**.

Complete these related subtraction facts.

1. 6 + 7 = 13   13 – ____ = 7   13 – 7 = ____
2. 8 + 9 = ____   ____ – 8 = ____   ____ – 9 = ____
3. 18 + 6 = ____   ____ – 18 = ____   ____ – 6 = ____
4. 14 + 7 = ____   ____ – 14 = ____   ____ – 7 = ____
5. 9 + 16 = ____   ____ – 9 = ____   ____ – 16 = ____
6. 34 + 11 = ____   ____ – 11 = ____   ____ – 34 = ____
7. 73 + 9 = ____   ____ – 73 = ____   ____ – 9 = ____
8. 88 + 5 = ____   ____ – 88 = ____   ____ – 5 = ____
9. 96 + 6 = ____   ____ – 6 = ____   ____ – 96 = ____
10. 104 + 26 = ____   ____ – 26 = ____   ____ – 104 = ____

SCORE /10 0-4 5-8 9-10

Statistics & Probability

AC9M3ST02

Data Collection

This table shows the fruit brought to school by Class 3B this week.

Fruit	Mon.	Tue.	Wed.	Thu.	Fri.	Totals
Apples	6	4	5	5	8	
Bananas	3	6	2	4	4	
Pears	2	0	2	3	3	
Mandarins	5	7	8	3	6	
Totals						

1. On which day was the most fruit brought to school? ____________
2. On which day was the least fruit brought to school? ____________
3. Which fruit was there most of? ____________
4. Which fruit was there least of? ____________
5. On which days was the same amount of fruit brought to school? ____________ and ____________
6. Three more bananas were brought on one day than on another.
   The days were: ____________ and ____________.
7. How much more fruit was brought to school on Tuesday than on Thursday? ____ pieces of fruit
8. How many pieces of fruit were brought to school by Class 3B this week? ____

SCORE /8 0-3 4-6 7-8

Measurement & Space

AC9M3M01

Measurement Units

Match each measurement unit to an item and a measuring tool.

① centimetres ② grams ③ litres ④ millilitres ⑤ kilograms

Items

Measuring Tools

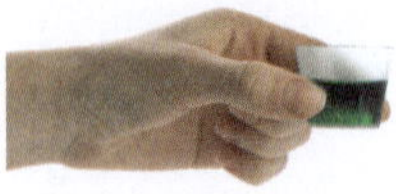

Estimate and name the measurement units for these items.

	Estimate	Units
⑥ The mass of an apple		
⑦ The length of a pencil		
⑧ A small bottle of juice		
⑨ The height of a giraffe		
⑩ A large bottle of milk		

SCORE /10 0-4 5-8 9-10

NAPLAN*-style Questions

① How many paper circles have been cut into quarters?

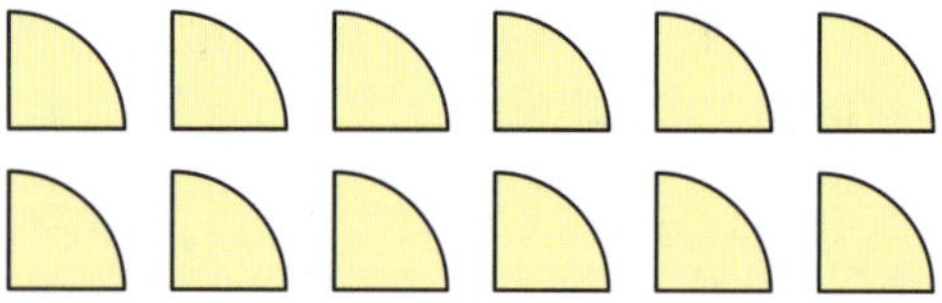

◯ 12 ◯ 6 ◯ 3 ◯ 2

② This table shows the mass of five kittens.
Which kitten is 300 g heavier than the lightest kitten?

____

Name	Ginger	Tabby	Tiger	Fluffy	Snowy
Mass	1030 g	970 g	1 kg	1230 g	730 g

③ How many blocks long is the pen?

◯ 5 blocks ◯ 7 blocks
◯ 3 blocks ◯ 4 blocks

SCORE /3 0-1 2 3

TARGETING MATHS HOMEWORK: YEAR 3 © PASCAL PRESS ISBN: 9781925726565

Problem of the Week

AC9M3N02

Fractions with Fish

"We are going to make a fraction fish," Mr Green told his class.
"We need to cut circles into fractions to make the scales, tail and fins," he said.

This fish has seven eighths and five quarters.

- Ada cut circles to make twelve quarters.
- Bindi cut three circles into fifths.
- Luis cut circles into sixteen eighths.
- Max cut five circles into thirds.

Question Time

1. How many circles did they cut up altogether? ____
2. Who cut up the most circles? ____________
3. Who made the most fraction pieces? ____________
4. How many fraction pieces did they make altogether? ____

Drawing a table can help you sort out the information in a problem like this.
Complete this table to help you answer questions 1–4.

	Circles	Fractions	Pieces
Ada			
Bindi			
Luis			
Max			
Total			

5. Did filling out the table help you answer the questions?

   ◯ Yes ◯ No

6. If so, would you try using a table for other problems?

   ◯ Yes ◯ No

Number & Algebra

AC9M3N01

Place Value

(1) (10) (100) (1000)

Answer these counter questions. The first one is done for you.

1. How many 10 counters make 100? 10
2. How many 10 counters make 1000? ______
3. How many 1 counters make 15? ______
4. How many 10 counters make 150? ______
5. How many 100 counters make 1500? ______
6. How many 1000 counters make 10 000? ______
7. How many 1 counters make 101? ______

Using only the counters shown, what is the smallest number of counters you need to make these numbers? The first one is done for you.

8. 101 — 0 (1000) 10 (10) 1 (1)
9. 1200 — ___ (100) ___ (10) ___ (1)
10. 1654 — ___ (1000) ___ (100) ___ (1)
11. 1236 — ___ (100) ___ (10) ___ (1)
12. 7063 — ___ (1000) ___ (100) ___ (1)
13. 8604 — ___ (1000) ___ (100) ___ (1)
14. 10 636 — ___ (1000) ___ (10) ___ (1)
15. 12 005 — ___ (1000) ___ (10) ___ (1)

SCORE /15 0-7 8-13 14-15

Quick Check

What number do these counters make?

1. 3 (100) 5 (1) ______
2. 6 (1000) 16 (10) ______
3. 17 (10) ______
4. 12 (10) 5 (1000) ______
5. 8 (1000) 13 (100) ______
6. 10 (100) 60 (1) ______

SCORE /6 0-2 3-4 5-6

TARGETING MATHS HOMEWORK: YEAR 3 © PASCAL PRESS ISBN: 9781925726565

Number & Algebra

UNIT 3

AC9M3A02

TERM 1

Bridging Back Through 10

We can find the difference between two numbers by bridging back through 10.

For 32 – 7, split the 7 into 2 and 5.

This is how we show it on a number line:

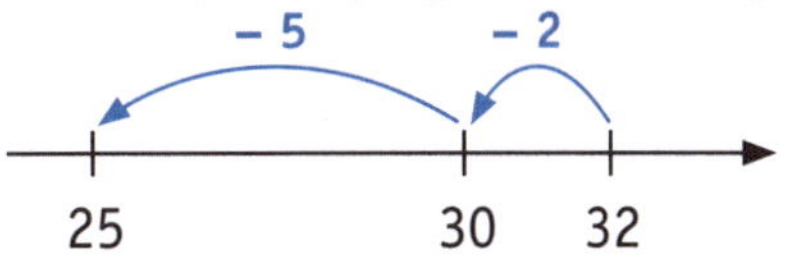

Show the bridging back through 10 on the number line. Then write the answer.

① 23 – 8 = _____

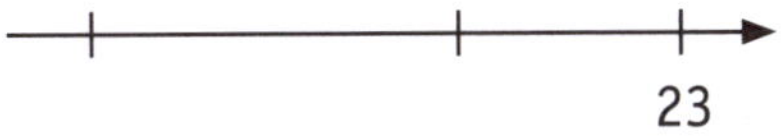

③ 42 – 6 = _____

② 44 – 6 = _____

④ 83 – 8 = _____

Use bridging back through 10 in your head for these subtractions.

⑤ 51 – 8 = _____

⑥ 26 – 9 = _____

⑦ 33 – 7 = _____

⑧ 62 – 7 = _____

SCORE /8 0-3 4-6 7-8

Statistics & Probability

AC9M3ST01

Max's Food Surveys

Max asked some friends, "Do you like broccoli?"
Here are his results:

Yes to broccoli	
No to broccoli	

Key:  = 2 people

① How many more people do not like broccoli than do like broccoli? _____

Max then asked some other friends about pears.
He found that 13 people didn't like pears and that 6 people did.

② Complete the graph to show Max's pear results.

Key:  = 2 people

SCORE /2 0 1 2

TARGETING MATHS HOMEWORK: YEAR 3 © PASCAL PRESS ISBN: 9781925726565

AC9M3SP01

3D Objects

Complete this table of the properties of some 3D objects.

3D objects	Faces	Edges	Vertices
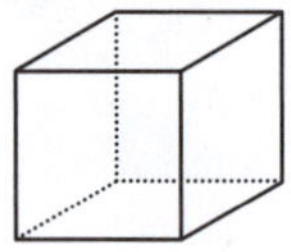			

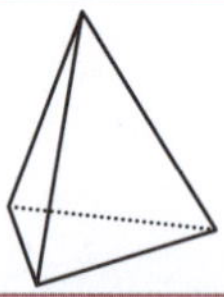			
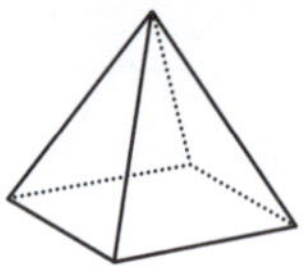			
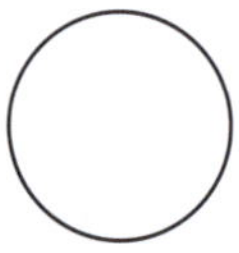			

SCORE /5 0-2 3-4 5

NAPLAN*-style Questions

1. Kiku saved $15 in 10c coins.
   How many 10c coins does she have?
   ______ coins

2. Look at the football crowd table.
   True or false?
   - Jerseys were the most popular clothing.
     ◯ True ◯ False
   - More people wore scarves than wore beanies.
     ◯ True ◯ False
   - There were 617 spectators counted.
     ◯ True ◯ False

The Football Crowd	
Clothing	**Number of spectators**
Beanies	165
Scarves	129
Club shirts	53
Jerseys	260

3. Which shapes were joined to make this model?
   ◯ A square-based pyramid and a cube
   ◯ Four cubes
   ◯ A rectangular prism and a cube

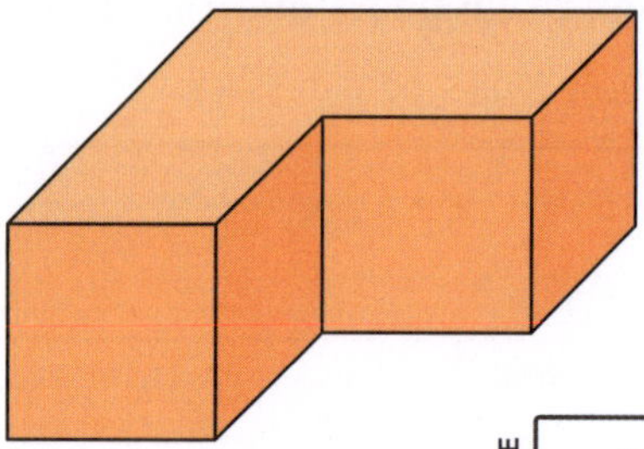

SCORE /3 0-1 2 3

TARGETING MATHS HOMEWORK: YEAR 3 © PASCAL PRESS ISBN: 9781925726565

Problem of the Week

AC9M3ST01

Data Representation

Hugo surveyed his friends about sports played after school.
He found that:

- 2 more people played football than played netball
- only 3 people played hockey
- twice as many played tennis as played hockey
- 1 more person played tennis than played netball
- each of his friends played only one sport after school.

Hugo's questions

1. How many people did Hugo survey? _____
2. How many more people played football than hockey? _____
3. How many people played tennis? _____

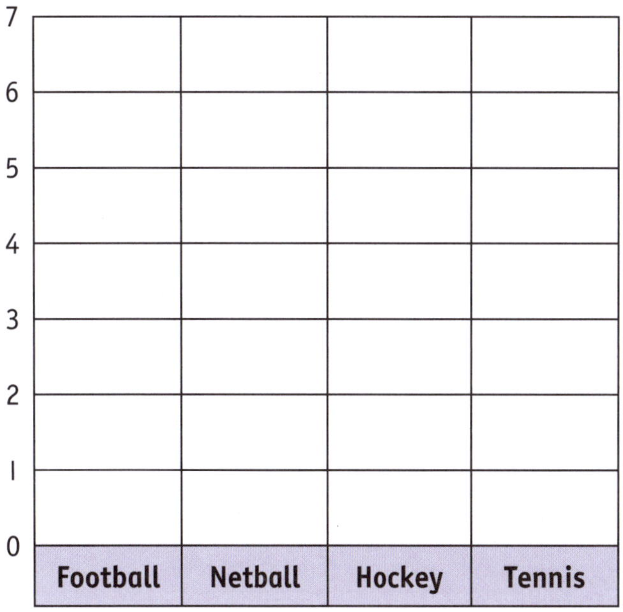

4. Which piece of information did you need to use first?

____

5. Write a question that could be answered by the chart of the data.

____

6. Did you enjoy sorting out this problem? ◯ Yes ◯ No

Why? ____

AC9M3N01

TERM 1

Place Value

What number is shown by the counters in these place-value charts? Write the answers in numerals and words.

(1)

Thousands	Hundreds	Tens	Ones
●	●	●●	●

Numerals: _______

In words: ____

(2)

Thousands	Hundreds	Tens	Ones
●●	●●		●

Numerals: _______

In words: ____

Using 4 counters on a similar place-value chart, record these numbers in numerals.

(3) The largest number possible _______

(4) The smallest number possible _______

(5) Three numbers with 0 in the 100s place _______, _______, _______

(6) Three numbers with 0 in the 10s place _______, _______, _______

(7) Three numbers with two digits the same in two places _______, _______, _______

(8) A number with no thousands _______

(9) A number in the tens counting pattern _______

(10) A number with two zeroes and two different digits _______

Write these numbers in words.

(11) 3405 ____

(12) 4316 ____

(13) 15 546 ____

(14) 14 039 ____

SCORE /14 0-6 7-12 13-14

Quick Check

Write these numbers as numerals.

(1) Five thousand and six _______

(2) Three thousand and seventeen _______

(3) Fifteen thousand, six hundred and seventy _______

(4) The number 1 less than 5000 _______

(5) The number 10 less than 15 000 _______

SCORE /5 0-2 3-4 5

TARGETING MATHS HOMEWORK: YEAR 3 © PASCAL PRESS ISBN: 9781925726565

Number & Algebra

AC9M3N06

Make to 10

There are six motorcycles, three trucks and four buses in the car park.
How many vehicles are there in the car park altogether?
Let's use the **make to 10** strategy.

$6 + 3 + 4 = 6 + 4 + 3 = 13$

Use make to 10 for these examples.

1. Five oranges, six bananas and five apples make

   ___ + ___ + ___ = ___ + ___ + ___ = ___ pieces of fruit altogether.

2. Three red roses, four yellow roses and seven pink roses make

   ___ + ___ + ___ = ___ + ___ + ___ = ___ roses altogether.

3. 8 + 5 + 2 = ___ + ___ + ___ = ___
4. 13 + 6 + 7 = ___ + ___ + ___ = ___
5. 26 + 5 + 4 = ___ + ___ + ___ = ___

SCORE /5 0-2 3-4 5

Statistics & Probability

AC9M3ST01

Bindi's Reading Chart

Bindi made a chart of the number of pages that she read last week.

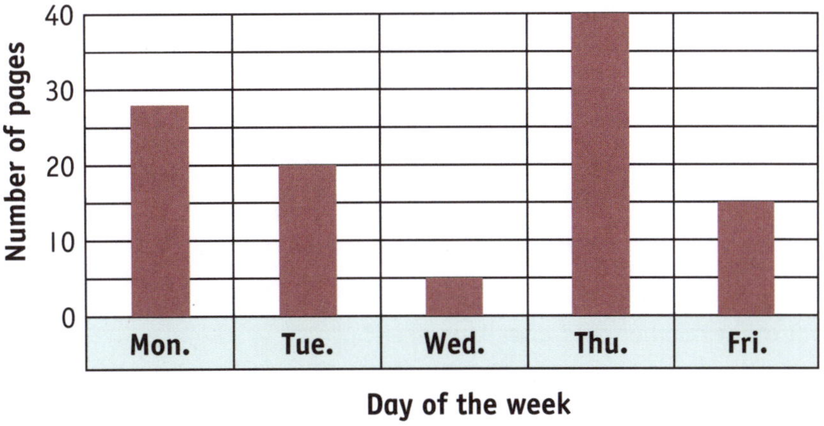

1. On which day did Bindi read the most? ________________
2. On which day did Bindi read 20 pages? ________________
3. How many pages did Bindi read on Friday? ____
4. How many pages did Bindi read altogether? ____
5. On which two days combined did Bindi read as much as she read on Tuesday?

   ________________ and ________________

SCORE /5 0-2 3-4 5

AC9M3M02

Measurement of Length

This row of blocks has been made with 1 cm blocks.

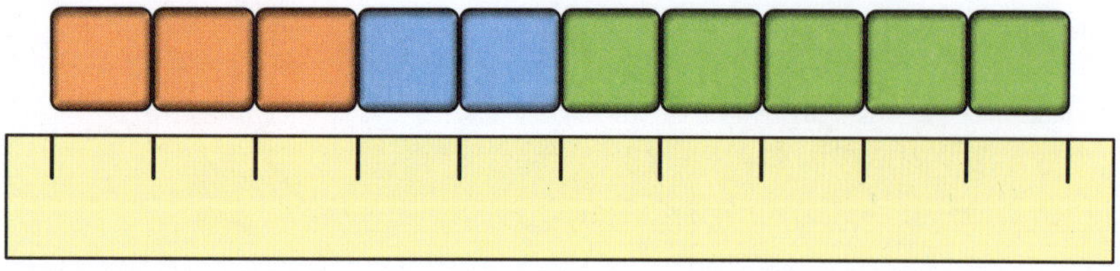

① How long is the section of orange blocks? ____ cm

② How long is the section of blue blocks? ____ cm

③ How long is the section of green blocks? ____ cm

④ How long is the row altogether? ____ cm

⑤ The ruler has been lined up below the blocks, but its numbers are missing. Add the numbers to the ruler.

Find 5 objects, estimate their length and check by measuring with a ruler.

	Object	Estimate	Actual length
⑥			
⑦			
⑧			
⑨			
⑩			

SCORE /10 0-4 5-8 9-10

NAPLAN*-style Questions

① Luis has these number cards: 3 6 7 8

To make the number thirty-eight thousand and sixty-seven, what other number card does Luis need? ☐

② Look at Max's Reading Log.

On how many days did Max read for less than 20 minutes?

____ days

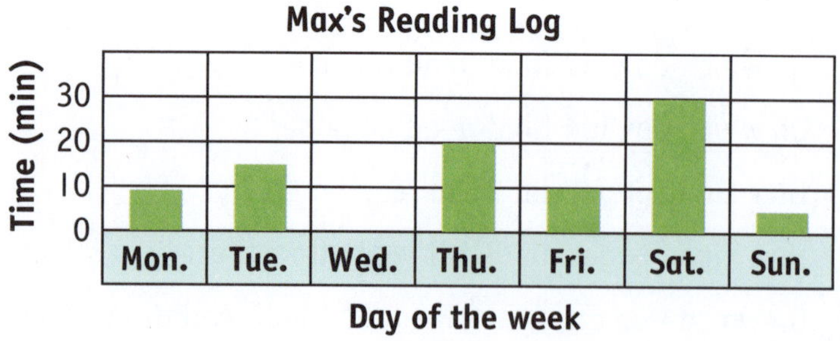

③ Kiku has a $5 note, six $1 coins and two $2 coins.

How much money does she have altogether?

- ◯ $5
- ◯ $8
- ◯ $13
- ◯ $15

SCORE /3 0-1 2 3

TARGETING MATHS HOMEWORK: YEAR 3 © PASCAL PRESS ISBN: 9781925726565

Problem of the Week

AC9M3ST01

Birds at the Bird Bath

Last week, Ada and Hugo collected information about the birds that visited their bird baths.

Ada collected this data:

Pigeons	𝍸 𝍸 ||
Finches	𝍸 𝍸 𝍸 |||
Sparrows	𝍸 𝍸 𝍸 𝍸 ||||
Magpies	𝍸 𝍸 𝍸 ||

Hugo collected this data:

Lorikeets	16
Blackbirds	22
Sparrows	10
Pigeons	24

Ada and Hugo want to show their data to their class using a block graph, with a scale of one block to 4 birds.

Use this space to show what their block graph should look like.

Number of bird visitors

Birds

1. How many birds visited the bird baths last week? ______
2. Which type of bird was the most frequent visitor? ______________
3. How many fewer finches visited than pigeons? ______
4. How does drawing the graph help to understand the data and answer the questions?

____

____

____

AC9M3N01, AC9M3M06

TERM 1

Making Money Amounts

There are no one-cent coins in use anymore but in the olden days, one dollar could be made with 100 one-cent coins.

Work out the total value for each number of coins.

							Total Value
1	20						
2		10					
3			5	2			
4	1	1			1		
5	10			1			
6	1	1	1			2	
7	5	1	1	3	1	1	

How much is in each purse?

8 ________

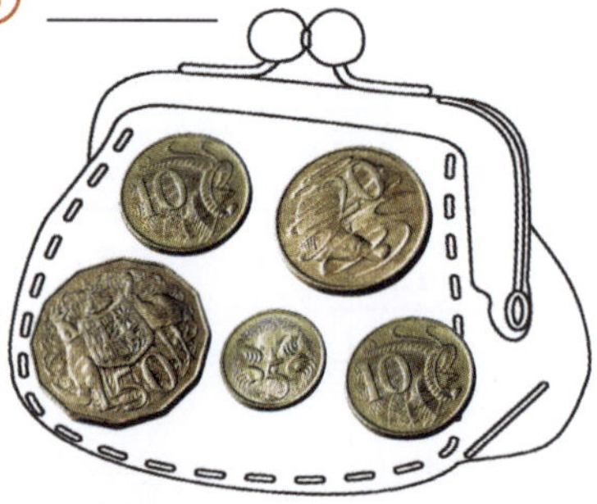

9 ________

10 ________

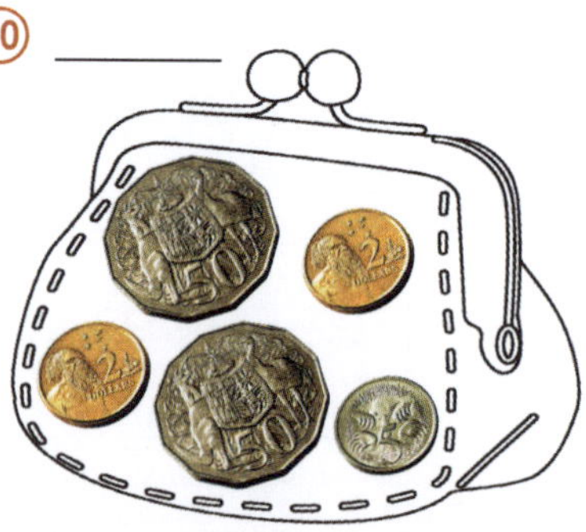

SCORE /10 0-4 5-8 9-10

Quick Check

Write the total amount for these coins.

1  ________

2  ________

3  ________

4  ________

5  ________

Which three coins can make these amounts?

6 60 cents ◯ ◯ ◯

7 $1·20 ◯ ◯ ◯

8 $2·10 ◯ ◯ ◯

9 $1·60 ◯ ◯ ◯

10 $3·05 ◯ ◯ ◯

SCORE /10 0-4 5-8 9-10

TARGETING MATHS HOMEWORK: YEAR 3 © PASCAL PRESS ISBN: 9781925726565

Number & Algebra

AC9M3N03

Adding by Partitioning

Partitioning or splitting numbers when you add makes **addition in the head** easier.
For 34 + 23, think like this:
34 + 20 is 54 and 3 more is 57.
It looks like this on a number line:

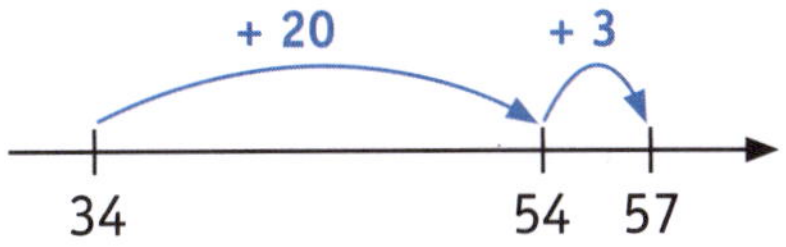

Show the partitioning on the number line. Then write the answer.

1. 27 + 12 = ____

2. 33 + 35 = ____

3. 46 + 43 = ____

Use partitioning in your head for these additions.

4. 17 + 12 = ____
5. 22 + 19 = ____
6. 13 + 13 = ____
7. 26 + 22 = ____
8. 43 + 34 = ____
9. 35 + 23 = ____
10. 17 + 11 = ____
11. 83 + 24 = ____
12. 75 + 26 = ____
13. 93 + 16 = ____
14. 76 + 34 = ____
15. 86 + 33 = ____

SCORE /15 0-7 8-13 14-15

Statistics & Probability

AC9M3P01

Coloured Counters

Look at the bag of counters.
How likely are you to choose these colours of counters from the bag? Tick one column to show your answer.

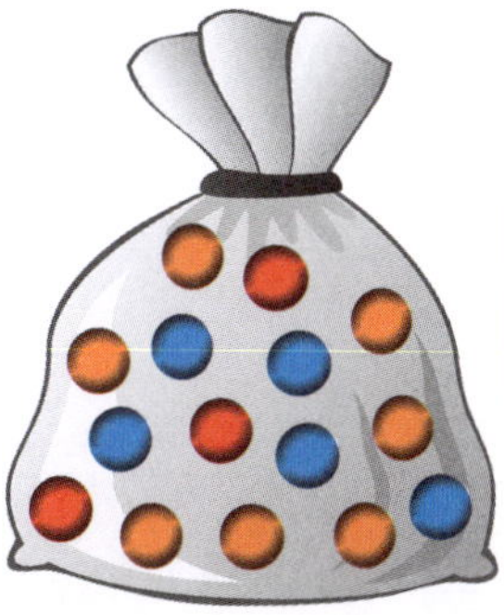

		Impossible	Quite likely	Very likely
1	Yellow			
2	Orange			
3	Red			
4	Pink			
5	Blue			

SCORE /5 0-2 3-4 5

Measurement & Space

AC9M3M01, AC9M3SP01

In the Art Room

① How long is the teacher's desk?

____

② How long is the art cupboard?

____

③ How long is the collage table?

____

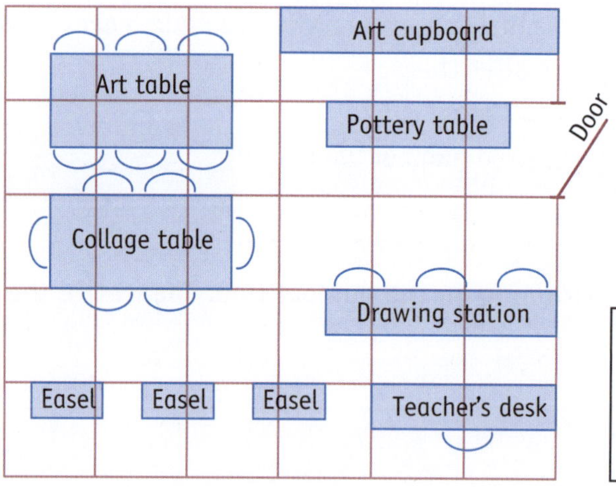

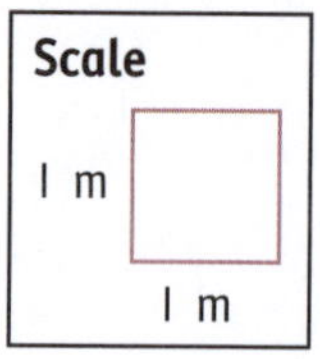

④ Complete these instructions for getting from the door to the easel in the corner.
Make sure that you don't have to squeeze past the collage table!

Go forward _____ m, turn ______ and go ahead _____ m.

Turn _______ and go ahead _____ m.

⑤ Write instructions to get from the back of the art table to the teacher's desk.

____

____

SCORE /5 0-2 3-4 5

NAPLAN*-style Questions

① Bindi has $1·05 in 5-cent coins.
How many 5-cent coins does she have?

☐ 5-cent coins

② Which building is furthest from the Office?

◯ Class 1 ◯ Class 3 ◯ Class 5
◯ Class 2 ◯ Class 4 ◯ Music
◯ Gym

School Plan

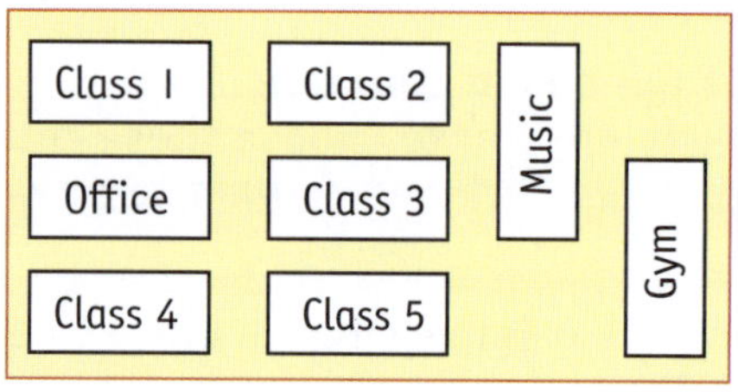

③ Max picked one of these counters.

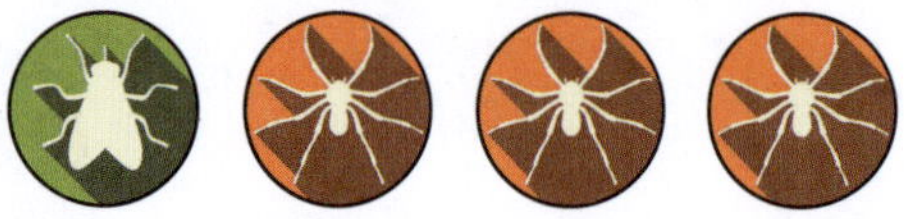

What is the chance that he picked a spider?

◯ Impossible ◯ Certain
◯ Unlikely ◯ Likely

SCORE /3 0-1 2 3

TARGETING MATHS HOMEWORK: YEAR 3 © PASCAL PRESS ISBN: 9781925726565

A New Reptile Park

The new reptile park will have glass panels so that visitors get a wonderful view of the animals and are safe from them too.

The zookeeper says:

- the crocodiles need the most space
- the snakes need twice as much space as the lizards
- the turtles need as much space as the snakes
- there has to be an open area for visitors to gather to view the animals.

The zoo has 21 large panels to use for the enclosures. Each glass panel is one grid-square long.

Use this grid to design the new reptile park.
Mark on your design which enclosure will house each of the reptile groups.

Feedback

Did using the grid help you design the reptile park?

◯ Yes ◯ No

If so, would you try using a grid for other problems?

◯ Yes ◯ No

AC9M3N07, AC9M3A03

Counting Sequences

Complete these counting sequences and related number sentences.

1. 0, 3, 6, 9, ____, ____, ____, ____, ____, ____, ____ 10 × 3 = ____
   How many 10s make 30? ____ × 10 = 30
2. 0, 4, 8, 12, ____, ____, ____, ____, ____, ____, ____ 10 × 4 = ____
   How many 10s make 40? ____ × 10 = 40
3. 0, 5, 10, 15, ____, ____, ____, ____, ____, ____, ____ 10 × 5 = ____
   How many 10s make 50? ____ × 10 = 50
4. 0, 10, 20, 30, ____, ____, ____, ____, ____, ____, ____ 10 × 10 = ____
   How many 10s make 100? ____ × 10 = 100
5. What do you notice about counting on 10 times in a number sequence?
   ____
6. Why do 3 × 10 and 10 × 3 give the same answer?
   ____

Use the pattern that you noticed to write the answers to these questions.

7. 10 × 1 = ____
   1 × 10 = ____
8. 10 × 2 = ____
   2 × 10 = ____
9. 10 × 6 = ____
   6 × 10 = ____
10. 10 × 9 = ____
    9 × 10 = ____

SCORE /10 0-4 5-8 9-10

Quick Check

How many times do you have to:

1. count on in 2s from 0 to get to 10? ____ times so ____ × 2 = 10
2. count on in 3s from 0 to get to 15? ____ times so ____ × 3 = 15
3. count on in 4s from 0 to get to 20? ____ times so ____ × 4 = 20
4. count on in 5s from 0 to get to 25? ____ times so ____ × 5 = 25
5. count on in 6s from 0 to get to 30? ____ times so ____ × 6 = 30

Answer these questions as quickly as you can. Use counting sequences to help if you need to!

6. 5 × 4 = ____
7. 6 × 3 = ____
8. 7 × 5 = ____
9. 7 × 10 = ____
10. 8 × 3 = ____

SCORE /10 0-4 5-8 9-10

TARGETING MATHS HOMEWORK: YEAR 3 © PASCAL PRESS ISBN: 9781925726565

Number & Algebra

AC9M3N03

Subtracting by Partitioning

Partitioning or splitting numbers when you subtract makes **subtraction in the head** easier.
For 34 – 23, split 23 into 20 and 3.
Then 34 – 20 is 14 and 14 – 3 is 11.
It looks like this on a number line:

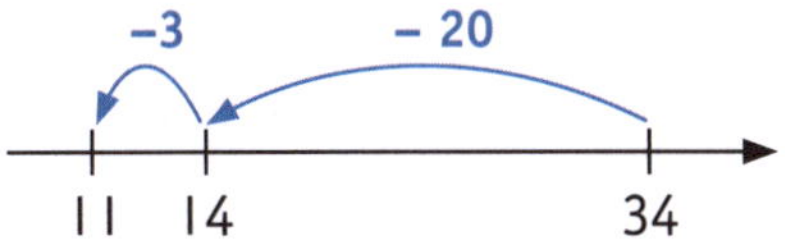

Show the partitioning on the number line. Then write the answer.

① 27 – 15 = ____

② 58 – 34 = ____

③ 46 – 23 = ____

Use partitioning in your head for these subtractions.

④ 67 – 14 = _____

⑤ 39 – 22 = _____

⑥ 54 – 23 = _____

⑦ 76 – 22 = _____

⑧ 87 – 14 = _____

⑨ 96 – 25 = _____

⑩ 85 – 11 = _____

⑪ 75 – 14 = _____

⑫ 96 – 33 = _____

⑬ 87 – 42 = _____

⑭ 69 – 34 = _____

⑮ 58 – 36 = _____

SCORE /15 0-7 8-13 14-15

Statistics & Probability

AC9M3P01

Food Survey

Imagine you surveyed 10 people about the popularity of some foods.
How likely are you to get each of these responses?
Rate the probability as certain, very likely, likely, unlikely, very unlikely or impossible.

	Survey response	Probability of response
①	I love snails and garlic.	
②	I love ice-cream and bananas.	
③	I don't mind broccoli on toast.	
④	I always have chilli on breakfast cereal.	
⑤	I don't mind orange juice.	

SCORE /5 0-2 3-4 5

AC9M3M05

TERM 1

Angles

Right angles are sometimes called **square angles** because at each corner of a square the angle is a right angle.

You can use the corner of this page to check if an angle is a right angle or if it is smaller or larger than a right angle.

Angles are everywhere!

Find 6 different angles in objects.
Use the corner of this book as an angle checker and complete this table.

	Object	Is it a right angle?	Is it larger than a right angle?	Is it smaller than a right angle?
1	Open door			
2				
3				
4				
5				
6				

SCORE /6 0-2 3-4 5-6

NAPLAN*-style Questions

1. Which coins could Hugo use to pay for this apple?

- ◯ 10c, 20c, 5c
- ◯ 50c, 50c, 20c, 5c
- ◯ $1, 20c, 50c, 5c
- ◯ 10c, 10c, 50c, 5c, 5c

2. Which angle is the largest?

3. Which colour is the spinner most likely to land on?

- ◯ Blue
- ◯ Red
- ◯ Green
- ◯ Yellow
- ◯ Orange

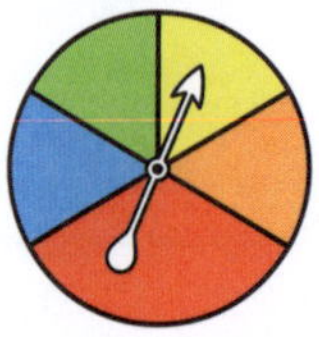

SCORE /3 0-1 2 3

TARGETING MATHS HOMEWORK: YEAR 3 © PASCAL PRESS ISBN: 9781925726565

Problem of the Week

AC9M3N03

As Close to Zero as Possible

Use these cards to make three 2-digit numbers.

0	1	2	3	4	5

Starting at 100, subtract each number to make your total as close to zero as possible.

For example, you could make the 2-digit numbers 45, 23 and 10.

$100 - 45 = 55$

$55 - 23 = 32$

$32 - 10 = 22$

That is quite close to 0, but can you get closer?

Questions

1. If you swap the digits of 45 to make 54, what will your final total be? _____
2. Can you use that idea to get even closer to zero?

Use this space to experiment.

Too easy? Try this one!

Use these cards to make three 2-digit numbers and one 1-digit number.

0	1	2	3	4	5	6

Starting at 100, subtract three of the numbers and add the fourth number to make your total as close to zero as possible.

Use this space to experiment.

TERM 1

Multiplication

To find a number in both the 2s counting pattern and the 5s counting pattern, tick the 2s facts and cross the 5s facts.

1 2✓ 3 4✓ 5✗ 6✓ 7 8✓ 9 10✗✓ 11 12✓ 13 14✓ 15✗ 16✓ 17 18✓

Record the facts like this: 2 jumps of 5 = 10, 2 × 5 = 10
5 jumps of 2 = 10, 5 × 2 = 10

① Find a number in both the 3s and 5s counting patterns.

1 2 3 4 5 6 7 8 9 10 11 12 13 14 15 16 17 18

___ jumps of ____ = ____, ____ × ____ = ____

___ jumps of ____ = ____, ____ × ____ = ____

② Find a number in both the 4s and 6s counting patterns.

1 2 3 4 5 6 7 8 9 10 11 12 13 14 15 16 17 18

___ jumps of ____ = ____, ____ × ____ = ____

___ jumps of ____ = ____, ____ × ____ = ____

4 20 6 12 9 30 8 7 40 5 13 25 10 15 18 16 49

Which numbers in the box above are in:

③ the 5s counting pattern? ____, ____, ____, ____, ____, ____, ____

④ the 4s counting pattern? ____, ____, ____, ____, ____, ____

⑤ the 3s counting pattern? ____, ____, ____, ____, ____, ____

⑥ the 7s counting pattern? ____, ____

⑦ the 10s counting pattern? ____, ____, ____, ____

⑧ Which number did not belong to any of those patterns? ____

SCORE /8 0-3 4-6 7-8

Quick Check

Write the answers to these multiplications.

① 3 × 2 = ☐

② 4 × 5 = ☐

③ 6 × 1 = ☐

④ 2 × 3 = ☐

⑤ 4 × 3 = ☐

⑥ 5 × 4 = ☐

⑦ 3 × 4 = ☐

⑧ 4 × 10 = ☐

⑨ 2 × 6 = ☐

⑩ 10 × 4 = ☐

⑪ 8 × 3 = ☐

⑫ 7 × 4 = ☐

⑬ 6 × 5 = ☐

⑭ 3 × 8 = ☐

SCORE /14 0-6 7-12 13-14

TARGETING MATHS HOMEWORK: YEAR 3 © PASCAL PRESS ISBN: 9781925726565

Number & Algebra

AC9M3A03

Multiplication Four Ways

Diagram	**Word Problem**
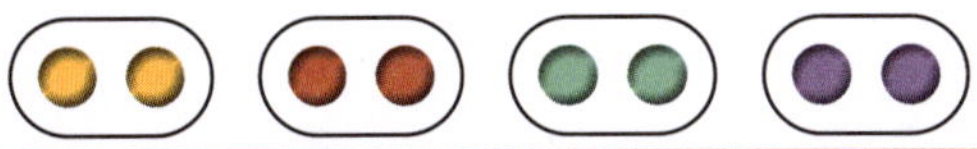	There are 4 plates and each plate has 2 apples. How many apples are there altogether?
Repeated Addition 2 + 2 + 2 + 2 = 8	**Multiplication Fact** 4 × 2 = 8

Complete the missing parts in these examples.

①	||| ||| ||| ___ + ___ + ___ = ___	3 groups of 3 tallies. How many tallies altogether?	___ × ___ = ___
②	______________________	4 groups of 4 tallies. How many tallies altogether?	___ × ___ = ___
③	______________________	6 groups of 5 tallies. How many tallies altogether?	___ × ___ = ___
④	||| ||| ||| ||| ||| ______________________	______________________ ______________________	___ × ___ = ___

Write the multiplication fact for these repeated additions.

⑤ 3 + 3 + 3 + 3 = ☐ × ☐ = ____

⑥ 2 + 2 + 2 = ☐ × ☐ = ____

⑦ 10 + 10 + 10 + 10 = ☐ × ☐ = ____

⑧ 5 + 5 + 5 + 5 + 5 = ☐ × ☐ = ____

⑨ 2 + 2 + 2 + 2 + 2 = ☐ × ☐ = ____

SCORE /9 0-4 5-7 8-9

Statistics & Probability

AC9M3P01

Will It Happen?

① Write three things that are certain to happen today.

______________ ______________ ______________

② Write two things that are impossible to happen today.

______________ ______________

③ Write something that is likely to happen in the next hour. ______________

④ Write something that is unlikely to happen in the next hour. ______________

⑤ Write something that is quite likely to happen today.

____

SCORE /5 0-2 3-4 5

AC9M3SP01

Making Models

These skeleton models of 3D objects were made with craft sticks and peas. Write the missing numbers.

1	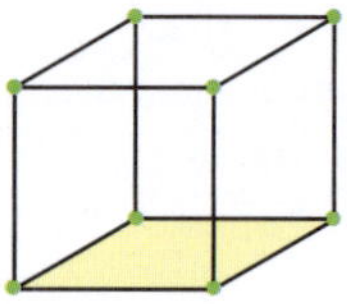	How many craft sticks were used? ____ How many peas were used? ____
2	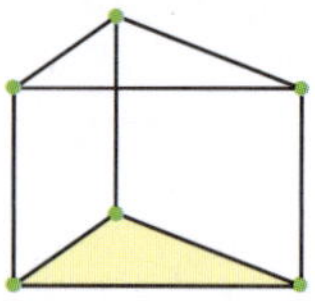	How many craft sticks were used? ____ How many peas were used? ____
3	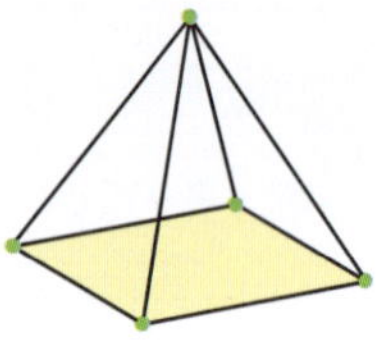	How many craft sticks were used? ____ How many peas were used? ____
4	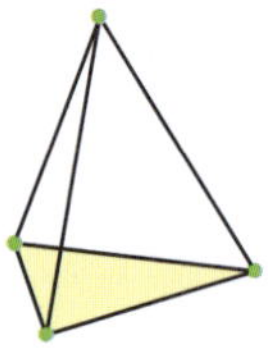	How many craft sticks were used? ____ How many peas were used? ____
5	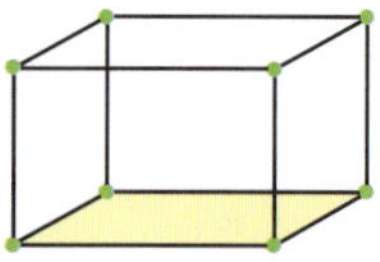	How many craft sticks were used? ____ How many peas were used? ____

SCORE /10 0-4 5-8 9-10

NAPLAN*-style Questions

1. The pet shop had four fish bowls with 5 fish in each bowl.
   How many fish altogether?
   - ◯ 4 ◯ 9 ◯ 5 ◯ 20

2. How many craft sticks were used to make the hexagonal pyramid model?
   - ◯ 6 ◯ 10 ◯ 12 ◯ 16

3. Luis put 5 black counters and 54 white counters in a bag.
   With looking, he picked out a counter from the bag, recorded its colour and then put it back.
   His colours were: white, black, white, white, white, black, black, black.
   What is likely to be the colour of the next counter that he picks?
   - ◯ Black ◯ White
   - ◯ Either black or white

SCORE /3 0-1 2 3

TARGETING MATHS HOMEWORK: YEAR 3 © PASCAL PRESS ISBN: 9781925726565

Problem of the Week

Target 100: A Multiplication Game

On their turn, players roll two dice.
They record their scores as a tally and a multiplication fact like this:

	𝍸 | 𝍸 |	2 × 6 = 12

At the end of Round 5, the player with the total score closest to 100 is the winner!

1. This is Max's score card for the first four rounds of the game.
Fill in the missing items.

Round 1		|| || || ||	☐ × ☐ = ☐
Round 2		𝍸 𝍸 𝍸 𝍸 𝍸 𝍸	☐ × ☐ = ☐
Round 3		|||| |||| |||| |||| ||||	☐ × ☐ = ☐
Round 4	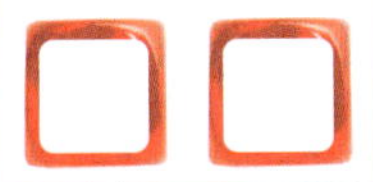	𝍸 | 𝍸 | 𝍸 |	☐ × ☐ = ☐

2. This is Kiku's score card for the first four rounds of the game.
Fill in the missing items.

Round 1		|||	☐ × ☐ = ☐
Round 2	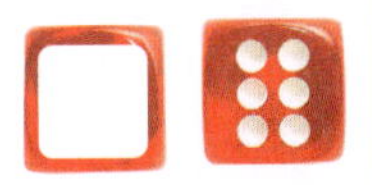	𝍸 | 𝍸 | 𝍸 | 𝍸 |	☐ × ☐ = ☐
Round 3		𝍸 𝍸 𝍸 𝍸 𝍸	☐ × ☐ = ☐
Round 4	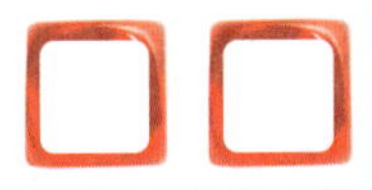	𝍸 | 𝍸 | 𝍸 | 𝍸 | 𝍸 | 𝍸 |	☐ × ☐ = ☐

3. Max hopes to score exactly 100. What does he need to score? ____
Which two dice give him that score? ______________________________

4. Kiku also hopes to score exactly 100. What does she need to score? ____
Which two dice give her that score? ______________________________

5. Who is more likely to score exactly 100? Why?

____

UNIT 8

Number & Algebra

AC9M3N02

Fractions of a Quantity

To find half of a quantity, make two equal-sized groups.

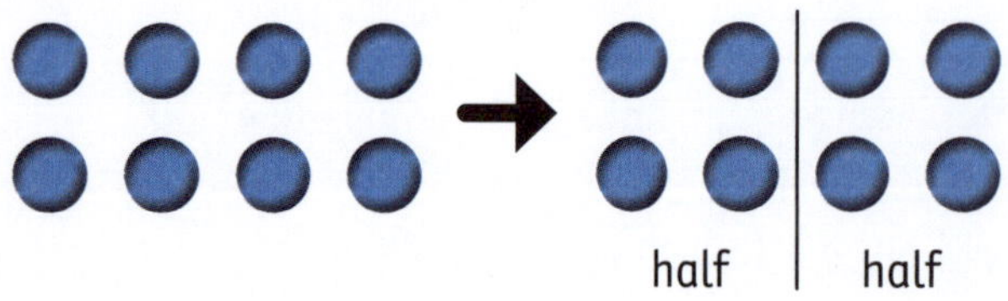

To find one-third of a group, make three equal-sized groups.

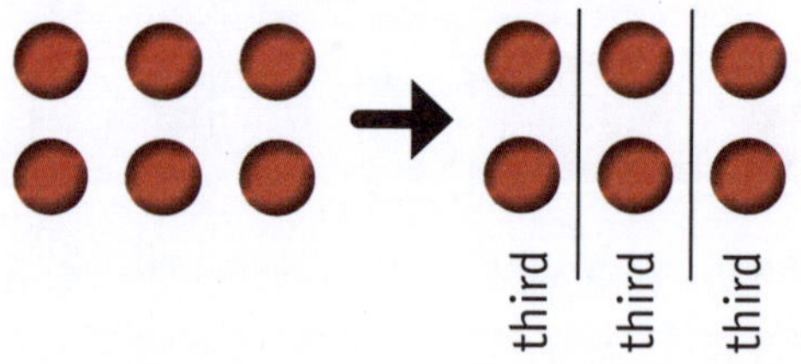

Show how to find these fractions of the groups shown. Then write the answer in the box.

(1) One-quarter of 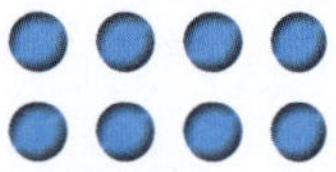 is ☐

(2) One-third of 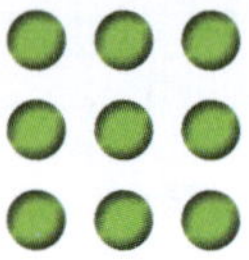 is ☐

(3) One-fifth of 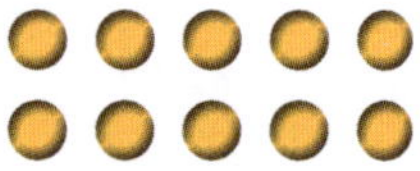 is ☐

(4) One-half of 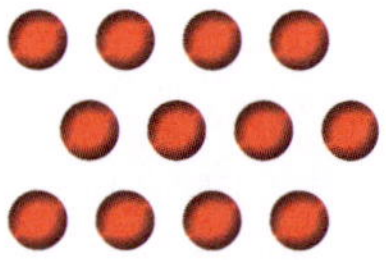 is ☐

(5) One-third of 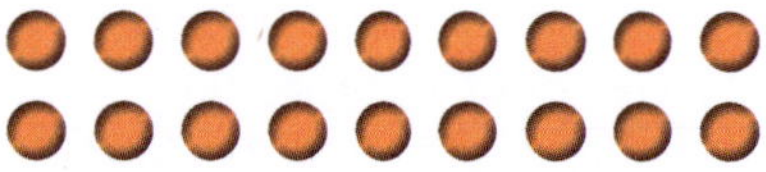 is ☐

(6) Three-quarters of 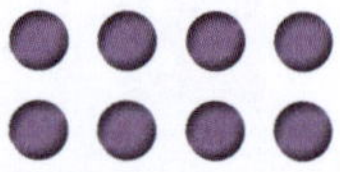 is ☐

(7) One-third of 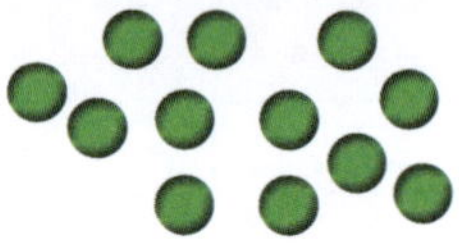 is ☐

(8) One-half of 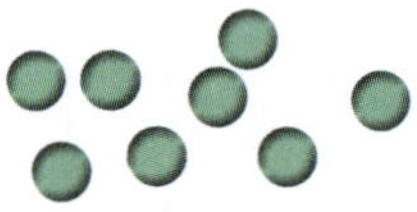 is ☐

(9) One-fifth of 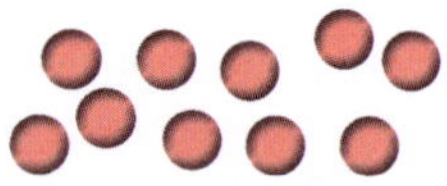 is ☐

(10) Two-thirds of 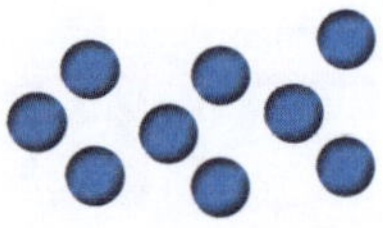 is ☐

SCORE /10 0-4 5-8 9-10

Quick Check

Work out these fractions of quantities. You can draw pictures to help.

(1) $\frac{1}{2}$ of 6 = ☐

(2) $\frac{1}{3}$ of 6 = ☐

(3) $\frac{1}{4}$ of 8 = ☐

(4) $\frac{1}{5}$ of 15 = ☐

(5) $\frac{1}{2}$ of 24 = ☐

(6) $\frac{1}{3}$ of 12 = ☐

(7) $\frac{1}{4}$ of 16 = ☐

(8) $\frac{1}{3}$ of 21 = ☐

(9) $\frac{1}{8}$ of 24 = ☐

(10) $\frac{1}{7}$ of 21 = ☐

SCORE /10 0-4 5-8 9-10

TARGETING MATHS HOMEWORK: YEAR 3 © PASCAL PRESS ISBN: 9781925726565

Number & Algebra

AC9M3N04

Multiplication Word Problems

Solve these word problems. You can draw diagrams to help.

1. Three muffins each had five blueberries on top.
   How many blueberries were there altogether? ____ blueberries
2. There were seven bunches of bananas with three in each bunch.
   How many bananas were there altogether? ____ bananas
3. Six pencils are $3 each.
   How much will I have to pay for all of them? ____ dollars
4. I gave my friend six bunches of poppies and there were ten poppies in each bunch.
   How many poppies was that altogether? ____ poppies
5. We have two hours of screen time every school day.
   How many hours is that in a whole school week? ____ hours

Multiplication Families

To show a multiplication family, draw a diagram and write the two ways of multiplying. The first one has been done for you. Complete the rest of the multiplication families.

6		2 × 3 3 + 3	3 × 2 2 + 2 + 2
7		____ ____	____ ____
8		3 × 5 5 + 5 + 5	____ ____
9		____ ____	____ ____
10	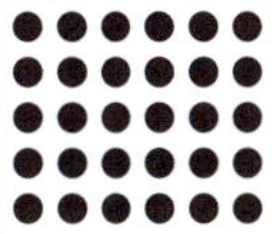	____ ____	____ ____

SCORE /10 0-4 5-8 9-10

Statistics & Probability

AC9M3ST02

School Travel Survey

Data Collected

Bus	Ada, Bindi, Luis, Max, Kiku, Hugo
Car	✓✓✓✓✓✓ ✓✓✓✓
Cycle	6
Walk	𝍸 𝍸 ||

Put the travel data on the graph in ascending order.

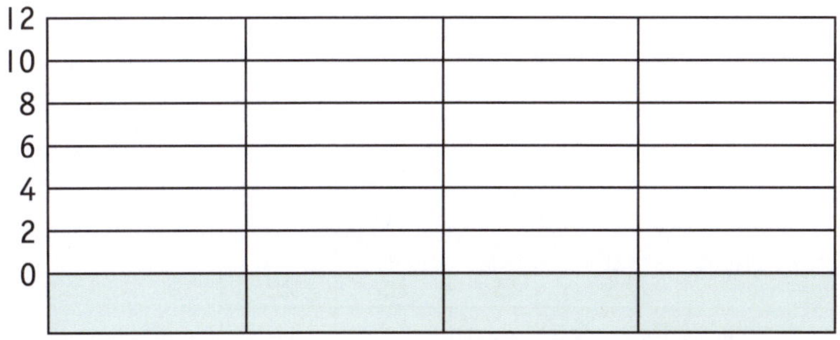

SCORE /4 0-1 2-3 4

AC9M3M03, AC9M3M04

Telling the Time

Where would the hour hand be pointing at these times? Draw the hour hands on the clocks.

1. 9 o'clock

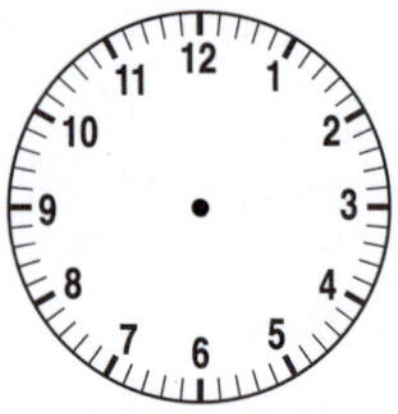

2. $\frac{1}{2}$ past 3

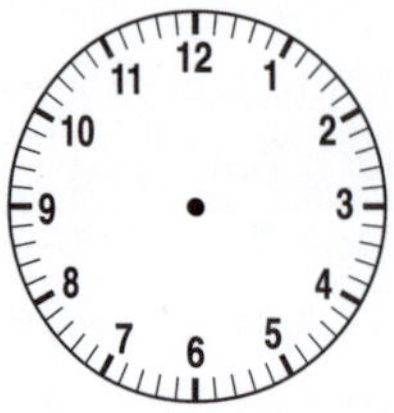

3. $\frac{1}{4}$ past 6

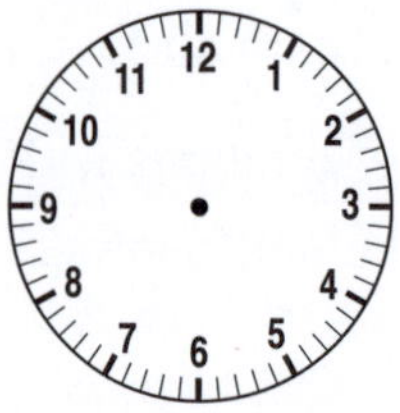

4. $\frac{1}{4}$ to 11

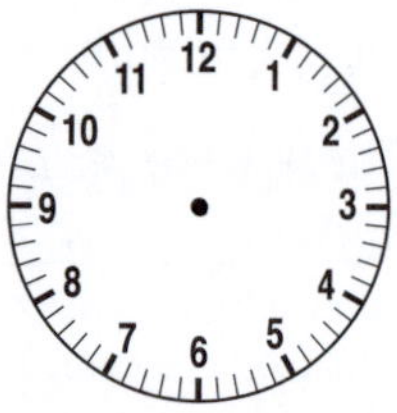

Where would the minute hand be pointing at these times? Draw the minute hands on the clocks.

5. 3 o'clock

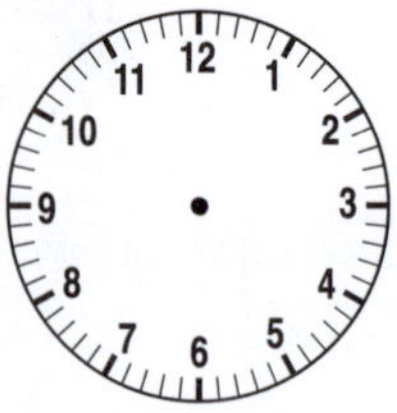

6. $\frac{1}{4}$ past 7

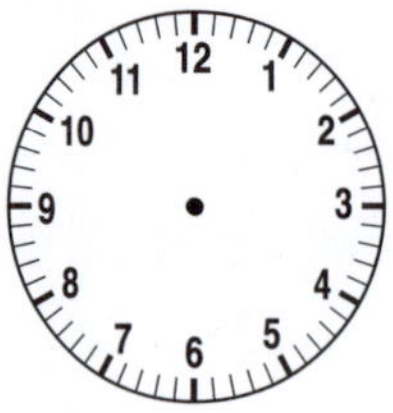

7. $\frac{1}{4}$ past 8

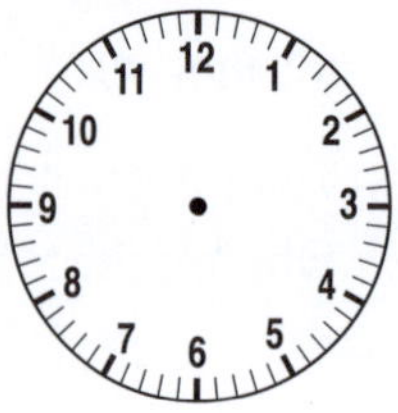

8. $\frac{1}{4}$ to 4

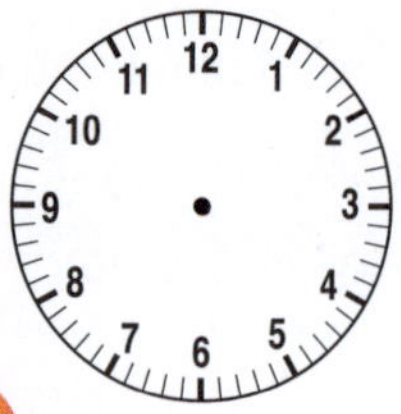

9. How many minutes in one hour? ______ minutes
10. How many hours in one day? ______ hours
11. How many seconds in one minute? ______ seconds
12. How many seconds in half a minute? ______ seconds
13. How many minutes in a quarter of an hour? ______ minutes

SCORE /13 0-6 7-11 12-13

NAPLAN*-style Questions

1. Ada shared 10 cherries with Bindi. How many cherries did Bindi get?

   Bindi got ☐ cherries.

2. Which clock shows half past 10?

   ◯ 
   ◯ 
   ◯ 
   ◯ 

3. This chart shows the 18 birds that visited my garden this morning.
   Complete the key for the graph.

Key  = __ birds

SCORE /3 0-1 2 3

TARGETING MATHS HOMEWORK: YEAR 3 © PASCAL PRESS ISBN: 9781925726565

Problem of the Week

AC9M3N02

Counters in the Bag

Luis, Max and Kiku each had a bag of 24 counters for their fraction problems.

1. **Luis's clues:**
   I have $\frac{1}{4}$ red, $\frac{1}{3}$ blue, $\frac{1}{6}$ green and the rest are yellow.
   How many yellow counters do I have? ____

2. **Max's clues:**
   I have $\frac{1}{3}$ yellow, $\frac{1}{4}$ blue and the same number of red as green.
   How many green counters do I have? ____

3. **Kiku's clues:**
   I have $\frac{1}{2}$ yellow and $\frac{1}{3}$ of what is left are green.
   I also have two more red than blue.
   How many blue counters do I have? ____

4. Who had the most blue counters?
   ____

5. Colour these counters to make your own bag full of counters.

6. Write fraction clues for your counter colours.
   My clues:
   ____
   ____
   ____

TERM 1 REVIEW

Review: Unit 1 and Unit 2

AC9M3N02, AC9M3N07, AC9M3A01, AC9M3A02, AC9M3M01, AC9M3SP02, AC9M3P01

TERM 1

Continue these sequences.

(1) 18, 20, ____, 24, ____, ____

(2) 19, 17, ____, ____, ____, 9

(3) 9, 12, ____, ____, 21, ____

(4) 101, ____, 97, 95, ____, ____

Write the number that is 1 more.

(5) 999 ____

(6) 1099 ____

(7) 1409 ____

Show the bridging through 10 on the number line. Then write the answer.

(8) 16 + 7 = ___

(9) 28 + 6 = ___

(10) Which colour is the spinner most likely to land on?

____

(11) Which colour is the spinner least likely to land on?

____

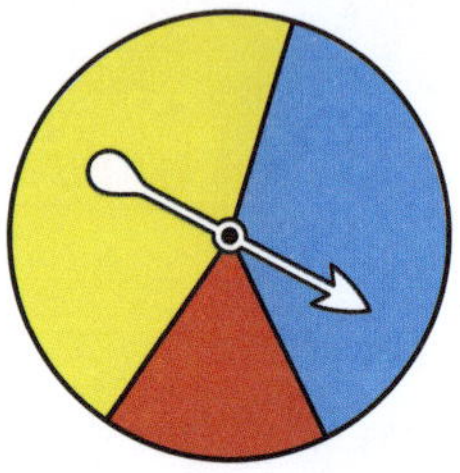

Draw the symbol that is:

(12) above the hash-tag and next to the tick. [ ]

(13) between the square and the cross. [ ]

(14) in the bottom left corner. [ ]

(15) two spaces above the semicircle. [ ]

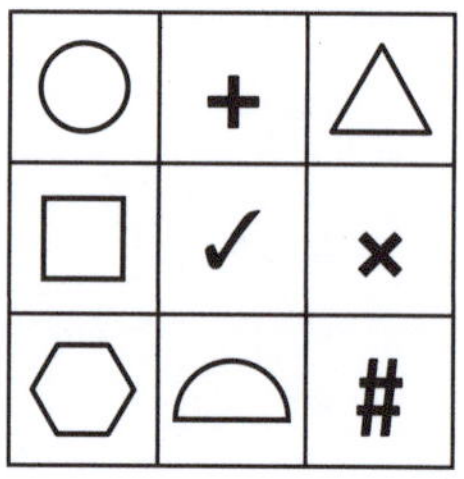

Write the fraction that is coloured blue.

(16) 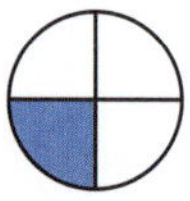 $\frac{\square}{\square}$

(17) 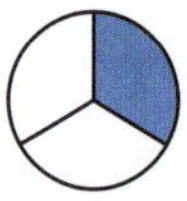 $\frac{\square}{\square}$

(18) 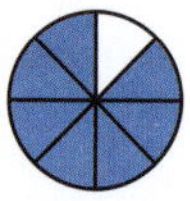 $\frac{\square}{\square}$

(19) If you know 9 + 4, what subtractions does that help with?

[ ] - [ ] = [ ] and [ ] - [ ] = [ ]

(20) Estimate the length of this line: ________________

Estimated length = ____ mm

Now measure to check.

Actual length = ____ mm

SCORE | /20 | 0-9 | 10-18 | 19-20

TARGETING MATHS HOMEWORK: YEAR 3 © PASCAL PRESS ISBN: 9781925726565

Review: Unit 3 and Unit 4

AC9M3N01, AC9M3A02, AC9M3ST01, AC9M3SP01

(1) How many 100 counters do you need to make 1100? ____ counters

(2) How many 10 counters do you need to make 130? ____ counters

What numbers do these counters make?

(3) 3 (1000) 4 (100) 2 (10) = ________

(4) 5 (1000) 11 (10) 6 (1) = ________

(5) 11 (100) 11 (10) 11 (1) = ________

Show the bridge back through 10 on the number line. Then write the answer.

(6) 33 − 7 = ____

(7) 52 − 6 = ____

(8) 94 − 8 = ____

(9) 103 − 7 = ____

The school soccer game was divided into four periods of 20 minutes with a 5-minute break between quarters.

The number of goals scored in each quarter is shown on the chart.

(10) How many goals were kicked altogether?

____ goals

(11) In which quarter were the most goals kicked?

____ quarter

(12) The total of the goals in which two quarters was the same as the number of goals in the first quarter?

________ quarter + ________ quarter = first quarter

Number of Goals Scored

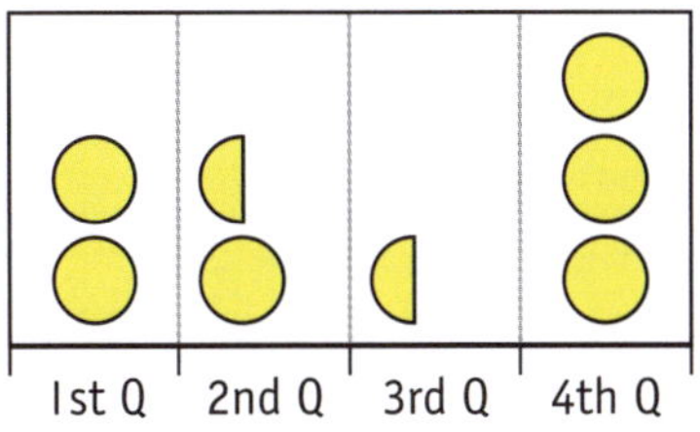

Key: ◯ = 2 goals

Write the correct letter in the box to match the 3D object to its description.

A = 5 faces and 6 vertices **B = 12 edges and 8 vertices** **C = 3 faces**

(13) 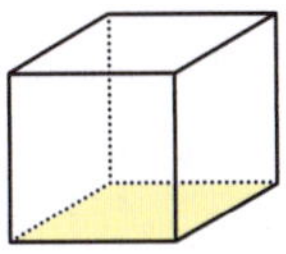 = ☐

(14)  = ☐

(15) 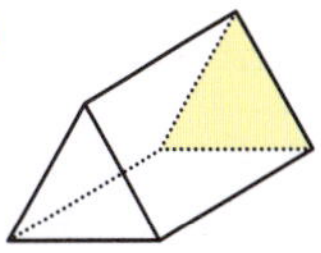 = ☐

SCORE /15 0-7 8-13 14-15

Review: Unit 5 and Unit 6

AC9M3N01, AC9M3N03, AC9M3N07, AC9M3A03, AC9M3M01, AC9M3M06, AC9M3SP01, AC9M3P01

How much money is in each purse?

① ______

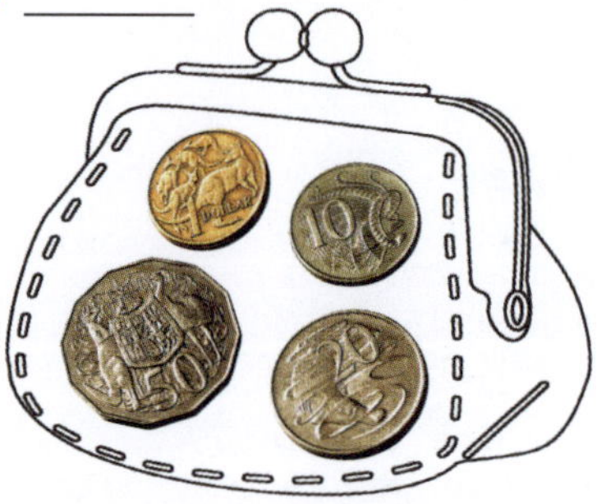

② ______

③ ______

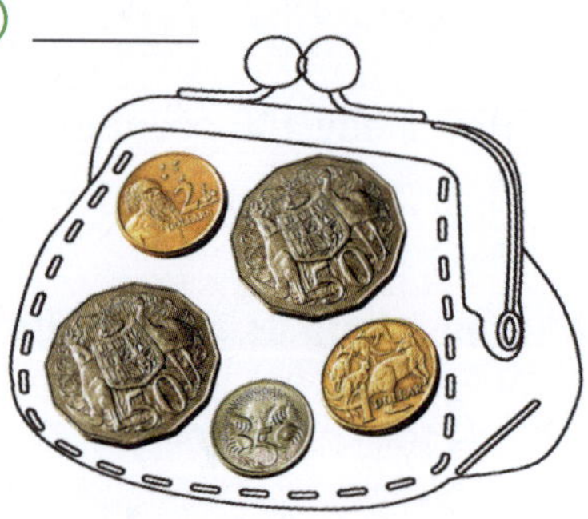

Show the partitioning on the number line. Then write the answer.

④ 55 − 23 = ____

⑤ 72 − 38 = ____

List exactly five moves that will get you from Red to Blue without crossing any of the blocked (grey) squares.

Use the words up, left, right or down to describe your moves and include the number of spaces to move.

⑥ Move 1: Go __________ ____ squares.

⑦ Move 2: Go __________ ____ squares.

⑧ Move 3: Go __________ ____ squares.

⑨ Move 4: Go __________ ____ squares.

⑩ Move 5: Go __________ ____ squares.

Complete these counting sequences and related number sentences.

⑪ 3, 6, 9, ____, ____, ____

☐ × ☐ = ☐

⑫ 4, 8, 12, ____, ____, ____

☐ × ☐ = ☐

⑬ How many times do you have to count in 5s to get to 45? ____ times

⑭ How many times do you have to count in 3s to get to 27? ____ times

⑮ Without looking, what is the chance of picking a 4 from these cards?

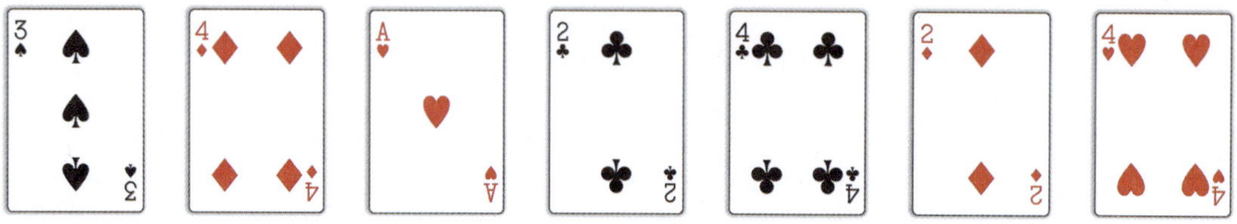

◯ Impossible ◯ Unlikely ◯ Likely ◯ Very likely ◯ Certain

SCORE	/15	0-7	8-13	14-15

TARGETING MATHS HOMEWORK: YEAR 3 © PASCAL PRESS ISBN: 9781925726565

Review: Unit 7 and Unit 8

AC9M3A03, AC9M3M03, AC9M3M04, AC9M3ST02, AC9M3SP01

Which numbers in the box are in:

1. the 5s counting pattern? ____________________
2. the 4s counting pattern? ____________________
3. the 3s counting pattern? ____________________

20 9 12 23 15 18 24

Complete these multiplications.

4. 3 × 4 = _____
5. 5 × 6 = _____
6. 7 × 3 = _____

Complete this 3D object table.

		Faces	Vertices	Edges
7	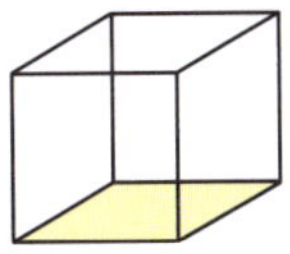			
8	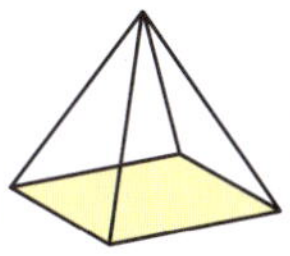			
9				

This graph shows how far Hugo, Bindi, Max and Ada live from school. Write the distances.

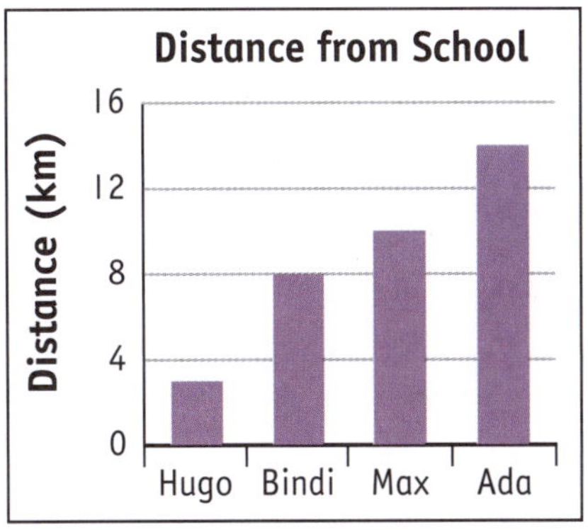

10. Hugo lives ____ km away from school.
11. Bindi lives ____ km away from school.
12. Max lives ____ km away from school.
13. Ada lives ____ km away from school.
14. What is their combined distance from school?

_____ km

Write the times in numbers and in words.

15. 

___:_____

____

16. 

___:_____

____

17. 

___:_____

____

SCORE /17 0-8 9-15 16-17

Estimation

① Look at the purple dots. Estimate but do not count how many purple dots there might be.

My estimate: ____ purple dots

② Draw on the diagram of purple dots to show an efficient way to find out how many purple dots there really are.

Actual number: ____ purple dots

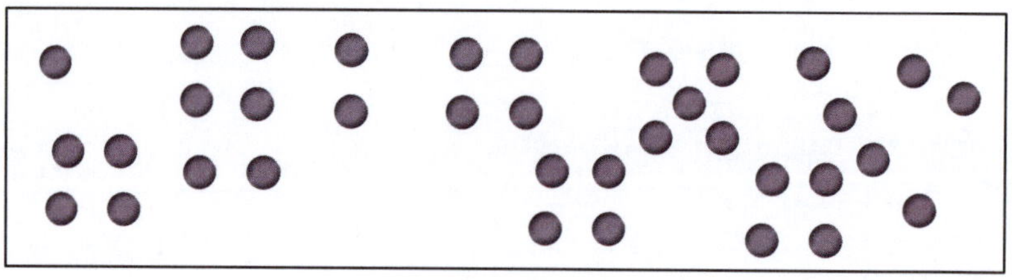

③ Estimate how many blue dots.

____ blue dots

④ Count to check.

Actual number: ____ blue dots

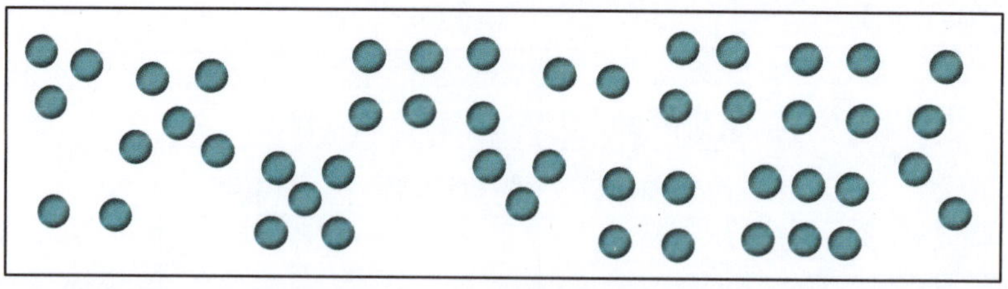

Using a Calculator

You can use a calculator to work out answers but you need to have a good idea of the expected answer first in case you press a wrong key.

For example, 38 + 48 will be more than 80 but less than 90, so a good estimate would be 85.

		Estimate	Answer	Good estimate?
⑤	74 – 56			◯ Yes ◯ No
⑥	3 × 19			◯ Yes ◯ No
⑦	76 + 78			◯ Yes ◯ No
⑧	48 ÷ 8			◯ Yes ◯ No
⑨	53 – 26			◯ Yes ◯ No
⑩	4 × 13			◯ Yes ◯ No

SCORE /10 0-4 5-8 9-10

Quick Check

Write estimates and then check.

		Estimate	Answer
①	66 – 37		
②	47 + 58		
③	5 × 22		
④	36 ÷ 3		

		Estimate	Answer
⑤	34 × 4		
⑥	4 × 16		
⑦	101 – 7		
⑧	15 + 10 + 12		

SCORE /8 0-3 4-6 7-8

TARGETING MATHS HOMEWORK: YEAR 3 © PASCAL PRESS ISBN: 9781925726565

Number & Algebra

Estimating Money Amounts

When estimating the totals of money amount, it is better to overestimate.

If you underestimate. you may find that you do not have enough money to pay with.

For example, if you estimate that $5·30 and $4·95 is about $9 and you have a $10 note, you will think you have enough to pay with.

In fact, $4·95 is almost $5 so your estimate should be over $10.

Try these estimation examples. The first one is done for you.

1. $6·50 + $7·20 is more than __$13__ but less than __$14__.
2. $3·80 + $4·50 is more than ________ but less than ________.
3. $12·75 + $8·50 is more than ________ but less than ________.
4. $7·20 + $7·50 is more than ________ but less than ________.
5. $29·60 + $32·70 is more than ________ but less than ________.

When shopping, you might want to buy more than two items.
Will $20 be enough for these items? Estimate, then tick Yes or No.

	Is $20 enough?	
6. $3·80 + $4·50 + $3·80	◯ Yes	◯ No
7. $3·95 + $5·95 + $9·50	◯ Yes	◯ No
8. $3·60 + $17·20 + 85c	◯ Yes	◯ No
9. $14·50 + $1·50 + $4·30	◯ Yes	◯ No
10. 25c + $4·60 + $13·75	◯ Yes	◯ No

SCORE /10 0-4 5-8 9-10

Statistics & Probability

AC9M3P02

Red or Black

This counter is red on one side and black on the other side.

If the counter is tossed 6 times, what outcomes could there be?

The first one is done for you.

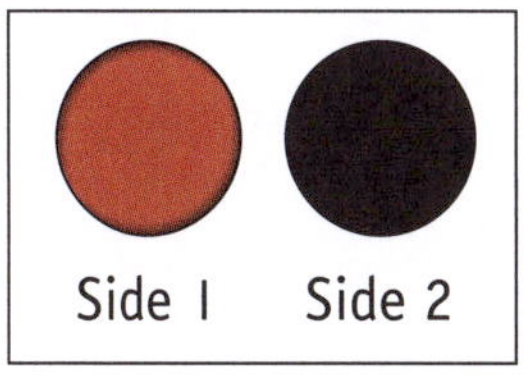

	6 tosses	
	Red	**Black**
1	3	3
2		
3		
4		
5		
6		
7		

8. Which outcome do you think is most likely?

____

9. Which outcomes do you think are least likely?

____________ and ____________

10. Why do you think that?

____

SCORE /10 0-4 5-8 9-10

Measurement & Space

ACM93SP02, ACM93M02

Turn and Turn Again

This is what happens when a bread tag makes a quarter turn to the right:

$\frac{1}{4}$ turn right

Draw in the box what the bread tag would look like after these turns.

① $\frac{1}{4}$ turn left

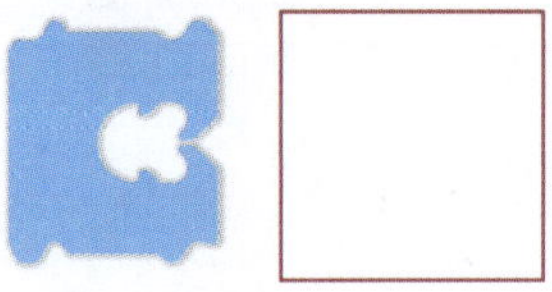

② $\frac{1}{2}$ turn

③ $\frac{3}{4}$ turn right

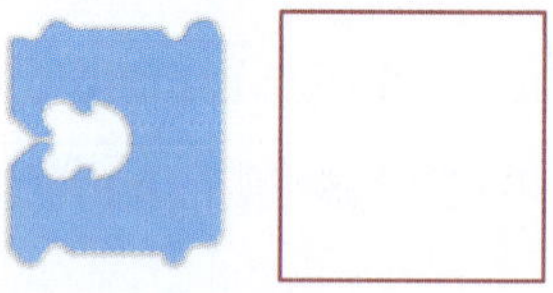

What will these clockfaces look like:

④ in half an hour?

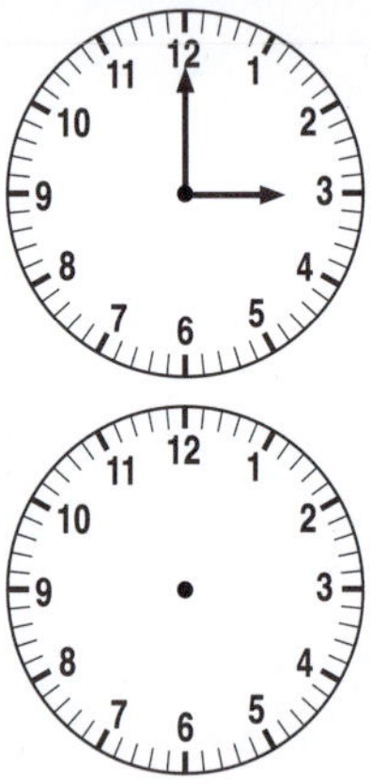

⑤ in $\frac{3}{4}$ of an hour?

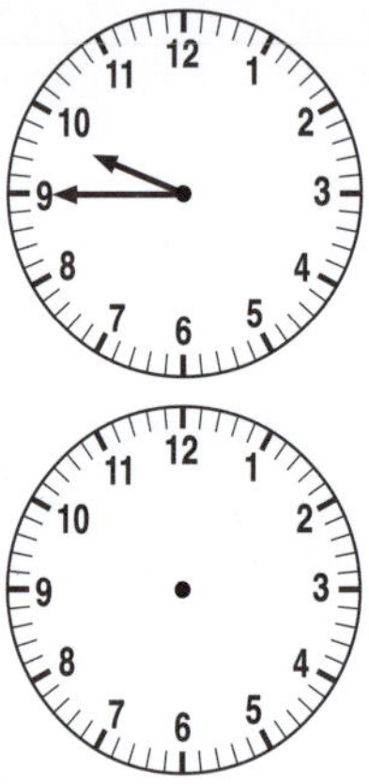

⑥ in $1\frac{1}{4}$ hours?

⑦ $\frac{1}{2}$ kilogram is ______ grams.

⑧ $\frac{1}{2}$ metre is ______ centimetres.

⑨ $\frac{1}{4}$ kilogram is ______ grams.

⑩ $\frac{1}{4}$ metre is ______ centimetres.

SCORE /10 0-4 5-8 9-10

NAPLAN*-style Questions

① The mass of four bananas is 1 kg. What is the approximate mass of 1 banana?

◯ 250 cm ◯ 25 g ◯ 100 g ◯ 250 g

② Luis has read between 17 and 30 pages of his book each day for the last three days. What is a reasonable estimate of how many pages he has read?

◯ 7 ◯ 60 ◯ 90 ◯ 30

③ This pattern is made by turning a card. Draw the next card.

SCORE /3 0-1 2 3

TARGETING MATHS HOMEWORK: YEAR 3 © PASCAL PRESS ISBN: 9781925726565

TERM 2

Choosing an End

Kiku is captain of the football team.

When the coin is tossed to decide which team will choose the starting end, she always wonders whether to call **heads** or **tails**.

Kiku decided to set up an experiment. She tossed a coin 10 times to see if heads or tails was more likely to win.

Toss	Heads	Tails
1	✓	
2	✓	
3		✓
4	✓	
5	✓	
6		✓
7		✓
8		✓
9	✓	
10	✓	

1. What was the result of Kiku's experiment?

   ______________ won more often.

Kiku thought she had nailed it.

But Max said, "Kiku, you have to do the experiment at least twice more to check if heads or tails has the better chance of winning."

Your Turn To Try

Toss a coin ten times and record the outcomes.

Then repeat the experiment.

2. What did your results tell you about the outcomes of heads and tails?

   ____

   ____

3. Are the results clear enough to show whether Kiku should call heads or tails?

   ____

4. Why or why not?

   ____

   ____

Experiment 1

Toss	Heads	Tails
1		
2		
3		
4		
5		
6		
7		
8		
9		
10		

Experiment 2

Toss	Heads	Tails
1		
2		
3		
4		
5		
6		
7		
8		
9		
10		

AC9M3N02

Fractions

The diagram shows that $\frac{1}{4}$ is shaded and $\frac{3}{4}$ is unshaded.

	Shaded	Unshaded	
	$\frac{1}{4}$	$\frac{3}{4}$	$\frac{1}{4}$ and $\frac{3}{4}$ = 1 whole

Complete the table to show the shaded and unshaded parts of each diagram.

		Shaded	Unshaded	
1	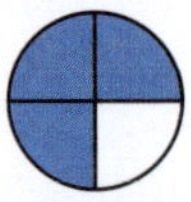	—	—	$\frac{3}{4}$ and — = 1 whole
2		—	—	— and — = 1 whole
3		—	—	— and — = 1 whole
4		—	—	— and — = 1 whole
5		—	—	— and — = 1 whole
6	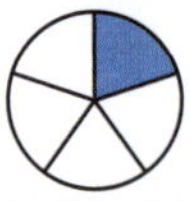	—	—	— and — = 1 whole

Colour these diagrams to match the description.

7  $\frac{2}{3}$ shaded

8  $\frac{3}{5}$ unshaded

9  $\frac{5}{8}$ shaded

10  $\frac{1}{6}$ unshaded

SCORE /10 0-4 5-8 9-10

Quick Check

What must the second fraction be to make one whole?

1. $\frac{1}{6}$ unshaded, — shaded
2. $\frac{5}{6}$ shaded, — unshaded
3. $\frac{2}{3}$ shaded, — unshaded
4. $\frac{7}{8}$ shaded, — unshaded
5. $\frac{3}{4}$ unshaded, — shaded
6. $\frac{3}{5}$ shaded, — unshaded
7. $\frac{3}{10}$ shaded, — unshaded
8. $\frac{2}{5}$ shaded, — unshaded
9. $\frac{4}{5}$ shaded, — unshaded
10. $\frac{2}{6}$ shaded, — unshaded

SCORE /10 0-4 5-8 9-10

TARGETING MATHS HOMEWORK: YEAR 3 © PASCAL PRESS ISBN: 9781925726565

Money Amounts

Show how to make half of each amount using the given number of coins.

1. Half of $4 using 3 coins ______ ______ ______
2. Half of $3 using 3 coins ______ ______ ______
3. Half of $1·40 using 2 coins ______ ______
4. Half of $7 using 4 coins ______ ______ ______ ______
5. Half of $1·90 using 4 coins ______ ______ ______ ______

Show how to make double each amount using the given number of coins.

6. Double $1·75 using 3 coins ______ ______ ______
7. Double $3·80 using 6 coins ______ ______ ______ ______ ______ ______
8. Double $1·95 using 6 coins ______ ______ ______ ______ ______ ______
9. Double $2·40 using 5 coins ______ ______ ______ ______ ______
10. Double $2·25 using the smallest possible number of coins ____________________

SCORE /10 (0-4) (5-8) (9-10)

Statistics & Probability

AC9M3P01

Changing Chances

How likely is each event? Tick the probability.

	Impossible	Unlikely	Likely	Very Likely	Certain
1. I will ride a camel today.					
2. I will ride a camel this year.					
3. I will go to bed tonight.					
4. I will go to bed every night.					
5. I will win a gold medal at the next Olympics.					
5. I will win a gold medal at an Olympics in the future.					

SCORE /5 (0-2) (3-4) (5)

AC9M3SP01

TERM 2

Making a Model

Bindi made this model using four different 3D objects.
Draw and name each 3D object that Bindi used.

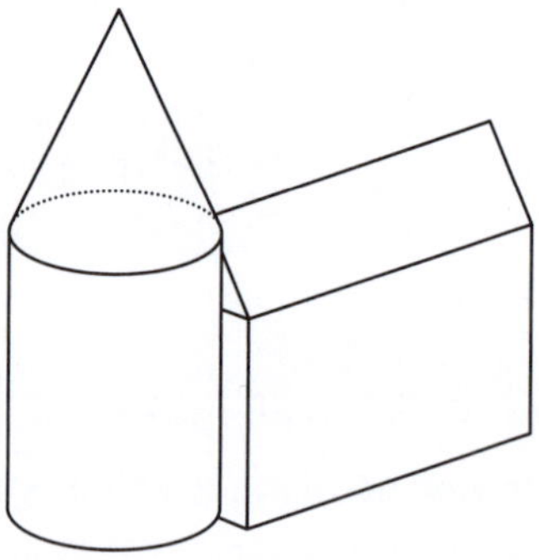

① ______________ ______________	③ ______________ ______________
② ______________ ______________	④ ______________ ______________

SCORE /10 0-4 5-8 9-10

NAPLAN*-style Questions

① Luis had these cherries.
He gave half of them to Max.
How many cherries did he give to Max?

⬭ 2 ⬭ 10 ⬭ 5 ⬭ 6

② Kiku had this money to share equally with Hugo.
How much should Hugo get?

$ [ ]

③ Ada built a tower using five objects.
Which of these objects could she not have used?

⬭ 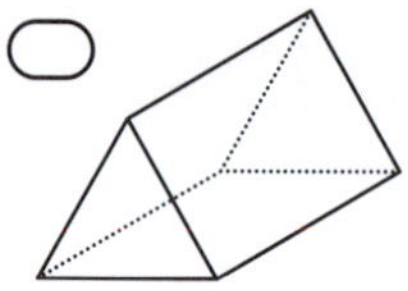 ⬭  ⬭ 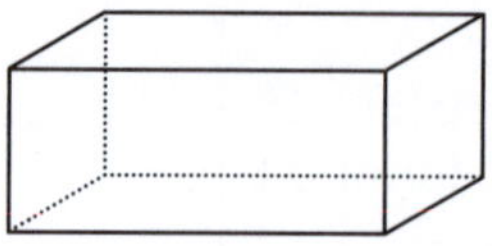 ⬭ 

SCORE /3 0-1 2 3

TARGETING MATHS HOMEWORK: YEAR 3 © PASCAL PRESS ISBN: 9781925726565

Equal Shares – No Leftovers

In the Baker family, biscuits have to be shared equally with no leftovers.
This family rule makes it hard sometimes to make fair shares.

You see, their favourite biscuits come in packets of 11, which is a terrible number for equal shares. The family has to cut fractions every time and that can be a very messy business.

1. With 11 biscuits in a packet, what would each member of a family of 2, 3, 4 or 5 get?

Family Size	Number of Biscuits
2	biscuits
3	biscuits
4	biscuits
5	biscuits

2. If there were 13 biscuits in a packet, what would each member of a family of 2, 3, 4 or 5 get?

Family Size	Number of Biscuits
2	biscuits
3	biscuits
4	biscuits
5	biscuits

3. What would be a good number of biscuits to have in a packet to make sharing easier? ____

4. With your chosen packet size, what would each member of a family of 2, 3, 4 or 5 get?

Family Size	Number of Biscuits
2	biscuits
3	biscuits
4	biscuits
5	biscuits

5. Did you draw biscuits to help with this problem?

◯ Yes ◯ No

6. If so, did drawing biscuits help?

◯ Yes ◯ No

Place Value

The same number can be written in different ways using place value.

For example, **364** can be written as:

36 tens and 4 ones
or 3 hundreds and 36 ones
or 35 tens and 14 ones.

2364 can be written as:

23 hundreds and 64 ones
or 236 tens and 4 ones
or 2 thousands, 35 tens and 14 ones.

Which number is different to the other two?

1. ◯ 1534
   ◯ 15 thousands and 34
   ◯ 15 hundreds, 3 tens and 4 ones
2. ◯ 6047
   ◯ 60 hundreds and 47 ones
   ◯ 60 thousands, 4 tens and 7 ones
3. ◯ 6506
   ◯ 65 hundreds and 6 ones
   ◯ 65 thousands and 6 ones
4. ◯ 11 110
   ◯ 11 hundreds and 1 tens
   ◯ 110 tens and 10 ones
5. ◯ 56 074
   ◯ 56 hundreds, 7 tens and 4 ones
   ◯ 56 thousands, 7 tens and 4 ones

Write these numbers in digits.

6. 36 hundreds, 5 tens and 6 ones ____________
7. 360 tens, and 6 ones ____________
8. 17 thousands, 5 tens and 5 ones ____________
9. 36 hundreds and 6 ones ____________
10. 170 tens and 5 ones ____________

SCORE /10 0-4 5-8 9-10

Quick Check

1. How many tens make 640? _____
2. How many hundreds make 6400? _____
3. How many hundreds make 10 000? _____
4. How many hundreds make 100 000? _____
5. How many tens make 1000? _____
6. 14 tens and 6 ones is _____.
7. 23 hundreds and 5 tens is _____.
8. 6 hundreds and 60 tens is _____.
9. 7 thousands and 11 hundreds is _____.
10. 12 thousands and 12 tens is _____.

SCORE /10 0-4 5-8 9-10

TARGETING MATHS HOMEWORK: YEAR 3 © PASCAL PRESS ISBN: 9781925726565

Number & Algebra

AC9M3N03

Partitioning for Addition

When adding numbers in the hundreds, it helps to work from the hundreds first.

For example, for **436 + 223**, partition into hundreds, tens and ones.

Hundreds: 400 + 200 = **600**   Tens: 30 + 20 = **50**   Ones: 6 + 3 = **9**

Then put the results together: 600 + 50 + 9 = **659**

Use partitioning for these additions.

		Hundreds	Tens	Ones	Total
1	275 + 123	200 + 100 = 300	70 + 20 = 90	5 + 3 = 8	300 + 90 + 8 = ______
2	183 + 216				
3	354 + 335				
4	124 + 173				
5	253 + 134				
6	726 + 273				
7	826 + 23				
8	133 + 175				
9	133 + 206				
10	563 + 524				

SCORE /10 0-4 5-8 9-10

Statistics & Probability

AC9M3ST01

School-night Bedtimes

A survey of Year 3 students gave the results shown in the graph.

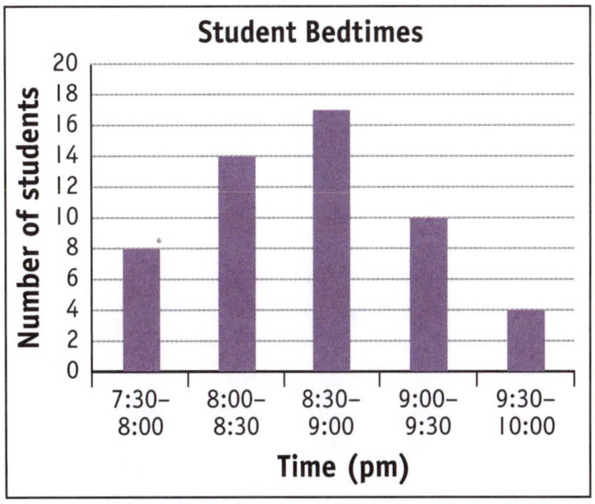

1. How many students were surveyed? ________
2. Which is the most common bedtime? ________
3. Which is the least common bedtime? ________
4. How many students go to bed between 8:00 and 9:00? ________
5. How many students go to bed after 9:00? ________

SCORE /5 0-2 3-4 5

AC9M3M03, AC9M3M04

About Time

The clocks show when the timers were set to start and finish. Complete the missing times. Then work out the difference between the Start Time and the Finish Time.

	Start Time	Finish Time	Duration
1	 :	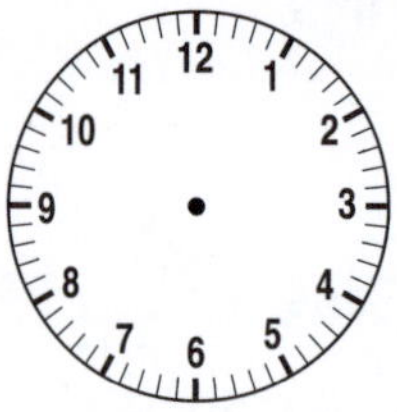 9:15	45 minutes
2	 :	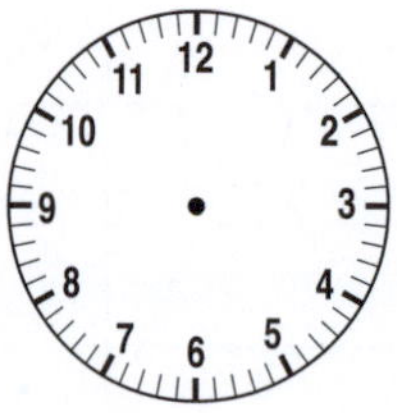 9:30	
3	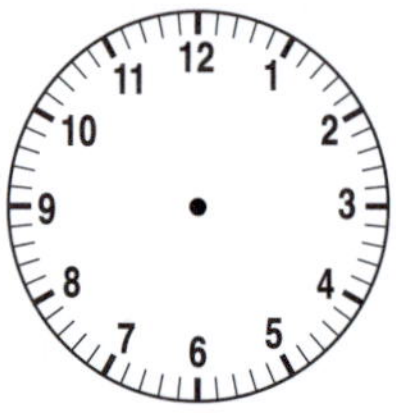 10:45	 :	

4 Which timer was set for the longest duration?

◯ 1 ◯ 2 ◯ 3

SCORE /10 0-4 5-8 9-10

NAPLAN*-style Questions

1 Hugo used this abacus to make a number. What is the number that he made?

◯ 2324 ◯ 23 024

◯ 20 324 ◯ 200 325

2 The next train leaves at:

What time is that in words?

____

3 How many pieces of rubbish were collected in the schoolyard?

____ pieces

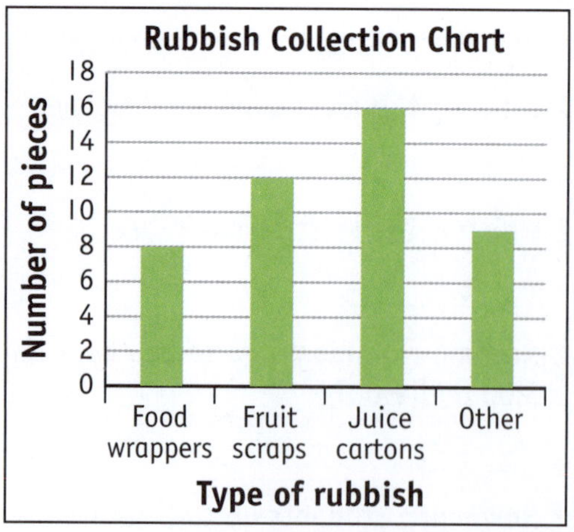

SCORE /3 0-1 2 3

TARGETING MATHS HOMEWORK: YEAR 3 © PASCAL PRESS ISBN: 9781925726565

Problem of the Week

AC9M3N03

The 600-page Readathon

Ada said, "I read 134 pages this week and 226 pages last week."

Max said, "I read 20 pages each night for 14 nights."

Bindi said, "I read 127 pages this week and 242 pages last week."

Kiku said, "I read twice as many pages as Max."

Luis said, "I read 243 pages last week and 72 pages this week."

Help Ada, Bindi, Luis, Max and Kiku make a reading plan for the next two weeks so that they will all reach their 600-page target by the end of Week 4.

	Last Week	This Week	Week 3	Week 4
Ada 				
Bindi 				
Luis 				
Max 				
Kiku 				

On an Empty Number Line

Follow the steps to work out where a given number fits on an empty number line.

Show where 12 lies on this number line.

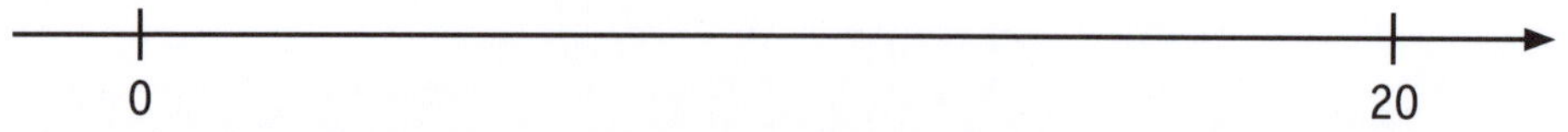

Step 1: Mark the point halfway between 0 and 20 (10).

Step 2: Mark the point halfway between 10 and 20 (15).

Step 3: Mark the point 12, which is slightly closer to 10 than 15.

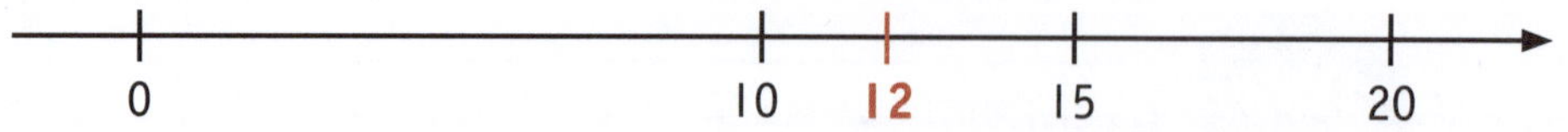

Work these examples.

1. Where is 13?

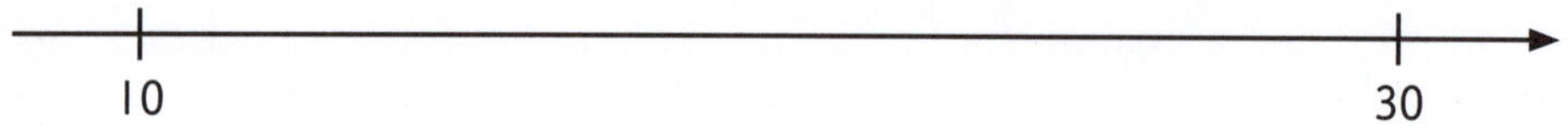

2. Where is 27?

3. Where is 96?

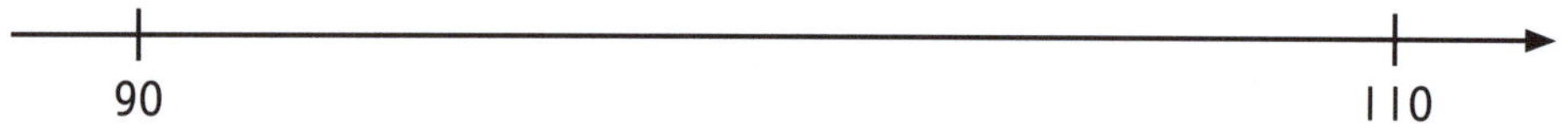

4. Where is 950?

5. Where is 24?

SCORE /5 | 0-2 | 3-4 | 5

Quick Check

What number is halfway between?

1. 6 and 18 ______
2. 22 and 44 ______
3. 15 and 23 ______
4. 100 and 150 ______
5. 120 and 230 ______
6. 8 and 16 ______
7. 12 and 24 ______
8. 600 and 800 ______
9. 72 and 82 ______
10. 20 and 50 ______
11. 90 and 110 ______
12. 131 and 151 ______
13. 1000 and 1500 ______
14. 9000 and 11 000 ______
15. 10 000 and 11 000 ______

SCORE /15 | 0-7 | 8-13 | 14-15

TARGETING MATHS HOMEWORK: YEAR 3 © PASCAL PRESS ISBN: 9781925726565

Number & Algebra

AC9M3N03

Partitioning for Subtraction

When subtracting numbers in the hundreds, it helps to work from the hundreds first.

For example, for **458 – 236**, partition into hundreds, tens and ones.

Hundreds: 400 – 200 = **200** Tens: 50 – 30 = **20** Ones: 8 – 6 = **2**

Then put the results together: 200 + 20 + 2 = **222**

Use partitioning for these examples.

		Hundreds	Tens	Ones	Total
1	275 – 123	200 – 100 = 100	70 – 20 = 50	5 – 3 = 2	100 + 50 + 2 = _____
2	359 – 238				
3	173 – 121				
4	726 – 315				
5	266 – 155				
6	568 – 342				
7	276 – 153				
8	678 – 454				
9	957 – 847				
10	867 – 344				

SCORE /10 0-4 5-8 9-10

Statistics & Probability

AC9M3P02

Spinners

1. If you spun the spinner 30 times, how often do you think it would land on each colour? Use tallies on the chart to show your answer.

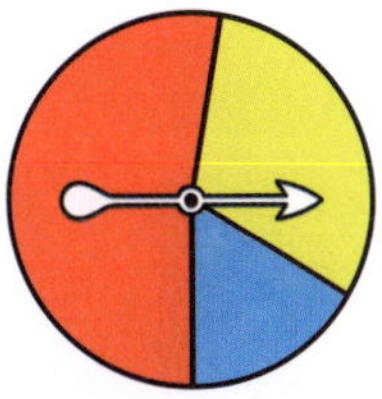

Tally Chart

Red	Blue	Yellow

2. Which colour do you think will be spun most often? ____________
3. Which colour do you think will be spun least often? ____________
4. Would the tally chart be the same for every 30 spins? ◯ Yes ◯ No
5. Why do you think that? ______________________________

____

SCORE /5 0-2 3-4 5

Turtle Moves

To move the turtle, the computer uses rules like:

Forward 2
Right 90°
Forward 2

The arrow shows the direction the turtle is facing.

Those moves will draw a diagram like this:

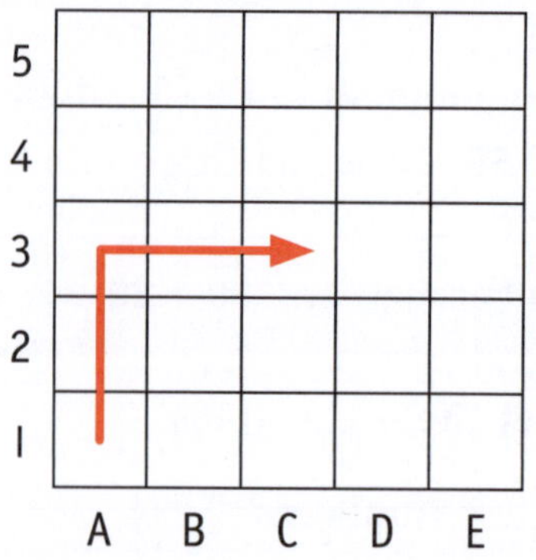

① What code would take the turtle from its position in C3 back to its starting point at A1?

________________ ________________

________________ ________________

Don't forget that the turtle is currently facing to the right.

② Now that the turtle is back at A1, what code will take it all the way around the grid, from A1 to A5 then on to E5 and E1 and finally back to A1?

________________ ________________

________________ ________________

________________ ________________

________________ ________________

Which way will the turtle be pointing when it starts this journey?

Not for Scoring

NAPLAN*-style Questions

① Write the missing number.

365 – ☐ = 123

② Which bag gives the least chance of picking a red counter?

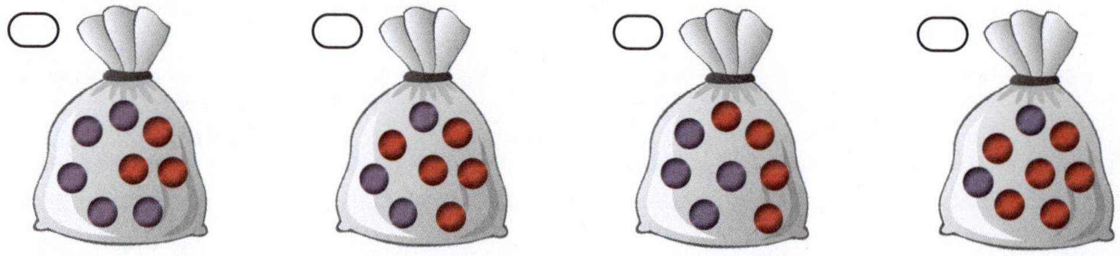

③ Which instructions will **not** move the turtle to position B1 on the grid?

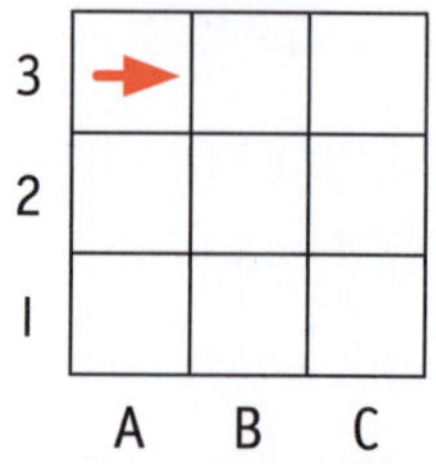

- ◯ Right 90°, Forward 2, Right 90°, Forward 1
- ◯ Right 90°, Forward 2, Left 90°, Forward 1
- ◯ Forward 2, Right 90°, Forward 2, Right 90°, Forward 1
- ◯ Forward 1, Right 90°, Forward 1, Forward 1

SCORE /3 0-1 2 3

TARGETING MATHS HOMEWORK: YEAR 3 © PASCAL PRESS ISBN: 9781925726565

Problem of the Week

Who Sat Where?

Luis, Max and Ada are friends.

Bindi, Kiku and Hugo are friends.

Use the clues to write the names in the correct places around the table.

Clues:
- Friends did not sit next to each other.
- Luis sat opposite Kiku.
- Max sat next to Hugo, on his left.

"I found it really helpful to write the names on small cards. I used two colours of cards to show who the friend groups were. That way I could move the cards around and check if all the clues were correct in my arrangements."

Did you use Mr Green's hint?

◯ Yes ◯ No

Did you find it useful to make the number cards?

◯ Yes ◯ No

AC9M3N01

TERM 2

Sequencing and Ordering of Large Numbers

These students each thought of a number and made up a clue to help you work out what the number is. The numbers are:

| 471 327 | 472 327 | 451 327 | 461 027 | 460 927 | 381 327 |

Write the correct number for each student.

1. 
Mine is the largest number. ________

4. My number has more in the 10 thousands position than any other number. ________

2. My number has fewer in the thousands place than any other number. ________

5. 
My number has zero in the hundreds position. ________

3. 
My number has fewer ten thousands than any other number. ________

6. My number is twenty thousand more than one of the other numbers. ________

SCORE /6 0-2 3-4 5-6

Quick Check

Are these statements true or false?

1. 472 360 has 7 in the ten thousands place.
◯ True ◯ False

2. 65 756 has 6 in the ten thousands place.
◯ True ◯ False

3. 650 650 has 0 in the thousands place.
◯ True ◯ False

4. 307 636 has 3 hundred thousands.
◯ True ◯ False

5. 457 607 has 5 in the thousands place.
◯ True ◯ False

6. 176 037 has 0 in the thousands place.
◯ True ◯ False

7. 375 809 is 37 thousand, 5 hundred and 89.
◯ True ◯ False

8. 567 999 is only 1 away from 568 000.
◯ True ◯ False

9. 407 637 is less than half a million.
◯ True ◯ False

10. 9976 is less than 1000.
◯ True ◯ False

SCORE /10 0-4 5-8 9-10

TARGETING MATHS HOMEWORK: YEAR 3 © PASCAL PRESS ISBN: 9781925726565

AC98M3N02

Extending Doubles Facts

If you know double 4, then you know double 40, double 400, double 4000 and so on. Complete these double tables.

	Start	4	40	400	4000	40 000
1	Double	8	80	800	8000	80 000
2	Double again	16	160	1600		
3	Double again	32	320			

The first two digits in each row stay the same – only their place value changes. Use this fact to complete the next two double tables.

	Start	6	60	600	6000	60 000
4	Double					
5	Double again					
6	Double again					

	Start	15	150	1500	15 000	150 000
7	Double					
8	Double again					
9	Double again					

SCORE /9 0-4 5-7 8-9

Statistics & Probability

AC9M3ST03

The Pizza Party

Bindi is planning a pizza party for her birthday. She gave her friends two votes.

"You have one vote for a pizza you like and one vote for a pizza you do not like," she told them.

Bindi made a chart to show the votes.

Pizza	Like	Dislike
Hawaiian	𝍸	𝍸
Meat Lovers	𝍸	𝍸 ||||
Pepperoni	||||	𝍸
Cheese and Tomato	𝍸 |||	𝍸
Supreme	𝍸 |	|||
Vegetarian	||	|||

Use Bindi's chart to answer the questions.

1. How many people are coming to the pizza party? ____
2. Which pizza was the most disliked? ________________
3. Which pizza was the least liked? ________________
4. Which pizza was liked as often as it was disliked? ________________
5. Which two pizzas should Bindi order for the party? ________________ and ________________
6. Why did you choose those two pizzas?

____

SCORE /6 0-2 3-4 5-6

AC9M3M01, AC9M3M02

Measuring Tools

The pictures show just some of the tools we use for measuring.

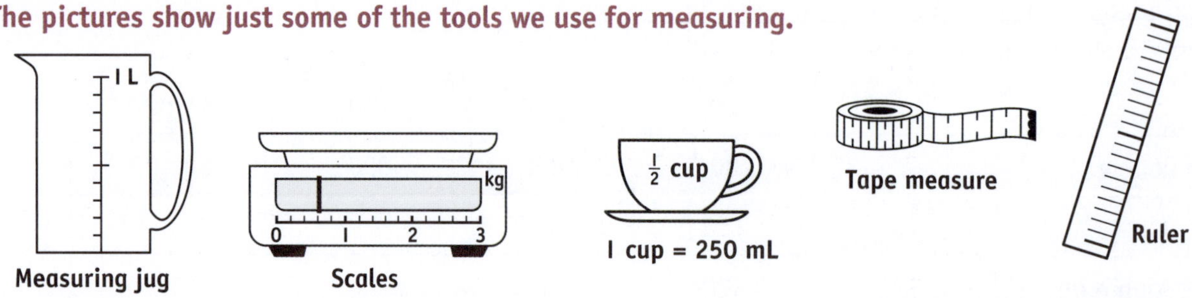

1. What measurement should be shown halfway up the scale on the measuring jug?

   ____

2. If the needle on the scales was halfway along the scale, what measurement would it be showing?

   ____

3. The marks on the ruler are 1 cm apart. What length of items can the ruler be used to measure?

   ____

4. I want to buy a new cap. Which tool should I use to measure around my head?

   ____

5. How many millilitres does the cup shown hold?

   ____

6. When might you use a cup for measuring?

   ____

7. My jam recipe needs $2\frac{1}{2}$ kg of sugar. Which tool should I use to measure it?

   ____

8. I need 1 litre of milk to make yoghurt. How many of the cups shown is that?

   ____

9. My book is shorter than the ruler but more than 15 cm. How long might it be?

   ____

10. What could you use the tape measure to measure?

    ____

SCORE /10 (0-4) (5-8) (9-10)

TERM 2

NAPLAN*-style Questions

1. Luis has 15 pens. Max has 6 more pens than Luis and Kiku has twice as many pens as Max.

   How many pens does Kiku have?

   ◯ 6 ◯ 15
   ◯ 21 ◯ 42

2. 12 students travelled to school by bus, 10 travelled by car and 4 walked to school.

   Write the correct numbers to complete the scale on the chart.

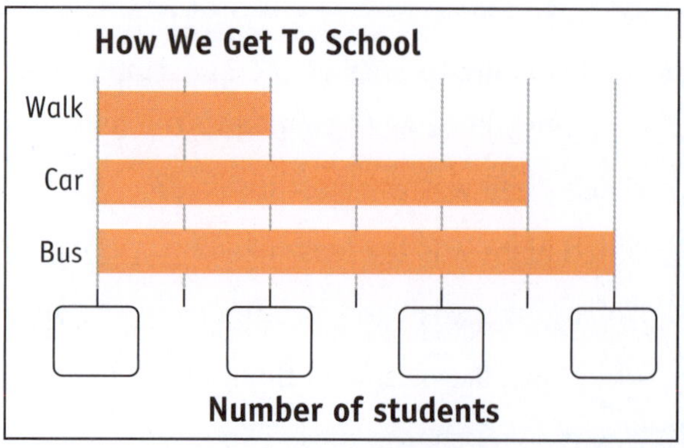

3. 

   What length is this straw closest to?

   ◯ 12 cm ◯ 3 cm
   ◯ 8 cm ◯ 7 cm

SCORE /3 (0-1) (2) (3)

TARGETING MATHS HOMEWORK: YEAR 3 © PASCAL PRESS ISBN: 9781925726565

Problem of the Week

AC9M3M02

Cordial Recipe

Luis is making a cordial drink for his friends.

Here is the recipe:

1 Add:

Orange cordial	100 mL
Lemon barley cordial	150 mL
Crushed ice	200 mL

2 Fill with iced water to 750 mL.

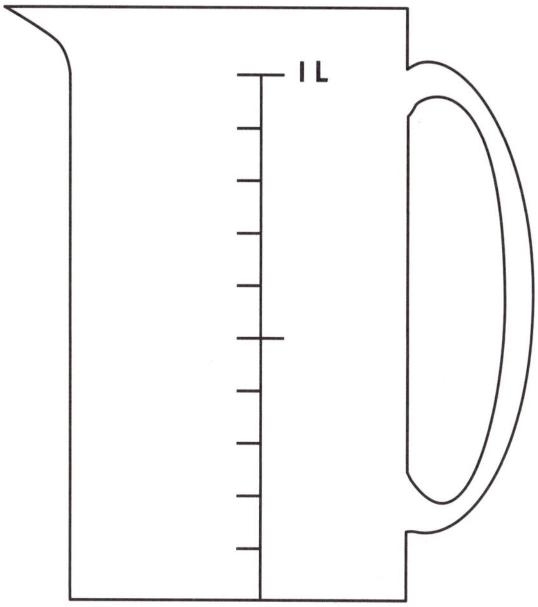

① How much iced water does Luis need to get from the fridge?

____

② Show the individual ingredients as layers in the jug.

③ Make up your own cordial recipe.

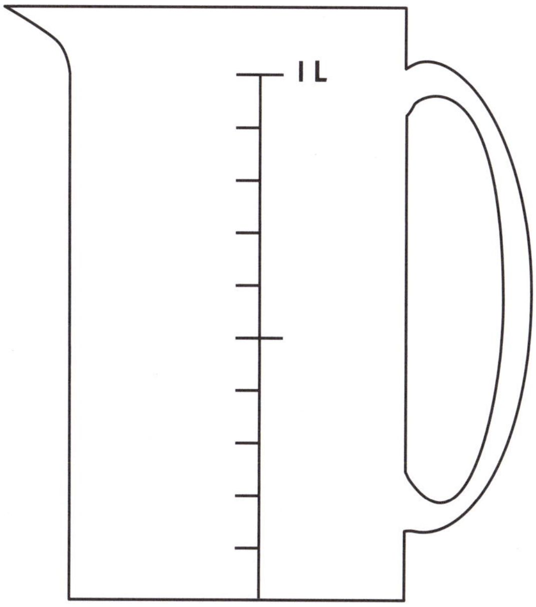

④ Show the individual ingredients as layers in the jug.

Number & Algebra

AC9M3N04, AC9M3A03

Multiplication with Tallies

You can use tallies to work out multiplication facts.

Let's use tallies to work out 7 × 6.

7 × 6 ⟨ 7 × 5 卌 卌 卌 卌 卌 卌 卌 = 35 + 7 × 1 | | | | | | | = 7 ⟩ **42**

Complete these tally multiplications.

① 4 × 8 ⟨ 4 × 5 卌 ____ ____ ____ = ___ + 4 × 3 ||| ____ ____ ____ = ___ ⟩ ___

② 6 × 7 ⟨ 6 × ___ ____ ____ ____ ____ ____ ____ = ___ + ___ × ___ ____ ____ ____ ____ ____ ____ = ___ ⟩ ___

③ 4 × 9 ⟨ ___ × ___ ____ ____ ____ ____ = ___ + ___ × ___ ____ ____ ____ ____ = ___ ⟩ ___

④ 6 × 8 ⟨ ___ × ___ ____ ____ ____ ____ ____ ____ = ___ + ___ × ___ ____ ____ ____ ____ ____ ____ = ___ ⟩ ___

You can also split the multiplications without using tallies:

6 × 7 = 6 × 5 + 6 × 2 = 30 + 12 = 42

Complete these multiplications.

⑤ 5 × 8 = ____________ = ________ = ____

⑥ 4 × 7 = ____________ = ________ = ____

⑦ 8 × 6 = ____________ = ________ = ____

⑧ 5 × 9 = ____________ = ________ = ____

SCORE /8 0-3 4-6 7-8

Quick Check

Multiply.

① 3 × 7 = ____

② 4 × 6 = ____

③ 3 × 9 = ____

④ 5 × 7 = ____

⑤ 7 × 6 = ____

⑥ 3 × 8 = ____

⑦ 5 × 9 = ____

⑧ 6 × 8 = ____

⑨ 3 × 6 = ____

⑩ 4 × 8 = ____

⑪ 3 × 11 = ____

⑫ 4 × 11 = ____

⑬ 5 × 11 = ____

⑭ 6 × 11 = ____

⑮ 7 × 11 = ____

SCORE /15 0-7 8-13 14-15

Number & Algebra

AC9M3N04

Multiplication Word Problems

To put 6 olives on each of 6 pizzas would need how many olives?

First write the multiplication number sentence: 6 × 6

Then work out the answer: 6 × 6 = **36**

Write the multiplication number sentences and then work out the answers.

1. 6 bunches of roses with 8 roses in each bunch. How many roses altogether?
   ☐ × ☐ = ☐
2. 5 oranges in each bag. How many oranges in 6 bags?
   ☐ × ☐ = ☐
3. 20 cents each for raffle tickets. How much for 8 raffle tickets?
   ☐ × ☐ = ☐
4. How many goldfish are needed to put 5 goldfish in each of 9 tanks?
   ☐ × ☐ = ☐
5. My dog eats 6 treats every day. How many treats does he eat in one week?
   ☐ × ☐ = ☐

SCORE /5 0-2 3-4 5

Statistics & Probability

AC9M3ST03

Favourite Movies

Ada is planning a movie evening with her friends.
She wants to make sure everyone has a good time.
She is wondering whether her friends prefer cartoons, animal movies or adventure movies.
She is also wondering whether there are some types of movies that her friends do not like.

Write three questions Ada could ask to find out the most and least popular types of movies among her friends.

1. ______________________________
   ____
2. ______________________________
   ____
3. ______________________________
   ____
4. Which of the questions you wrote would help Ada gather the information?
   ◯ Q1 ◯ Q2 ◯ Q3
5. Which question do you think would be least effective?
   ◯ Q1 ◯ Q2 ◯ Q3

SCORE /5 0-2 3-4 5

Paper Folding to Make Angles

Bindi folded her paper square twice and then opened it up. She drew along the lines of the folds, and this was the result.

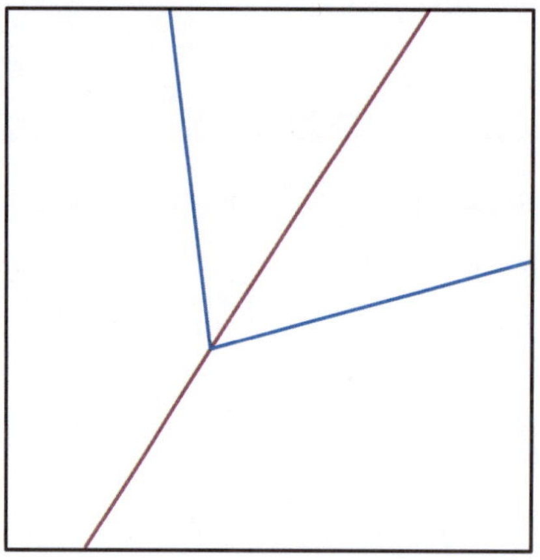

Bindi's Questions

1. How many angles are there altogether inside the paper square? _____
2. How many angles are right angles? _____
3. How many angles are larger than a right angle? _____
4. How many angles are smaller than a right angle? _____
5. How many pairs of angles the same size can you find? _____

SCORE /5 (0-2) (3-4) (5)

NAPLAN*-style Questions

1. Luis spent $5 every school day last week. How much did he spend altogether?

   ◯ $5 ◯ $35 ◯ $25 ◯ $12

2. The door was open more than halfway.
   What was the angle made between the door and the wall?

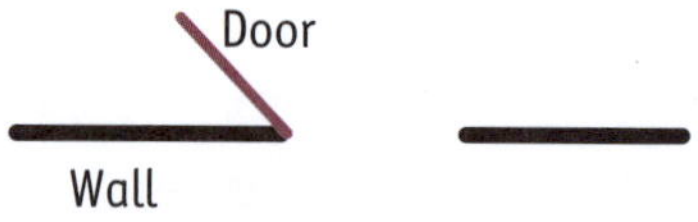

   ◯ Straight line angle
   ◯ Right angle
   ◯ More than a right angle
   ◯ Less than a right angle

3. How many points were scored altogether at the quiz night?

   ◯ 12 ◯ 40
   ◯ 42 ◯ 18

Quiz Night Scores	
Blue Team	卌 卌 |||
Red Team	卌 卌 ||
Green Team	卌 卌 卌 ||

SCORE /3 (0-1) (2) (3)

TARGETING MATHS HOMEWORK: YEAR 3 © PASCAL PRESS ISBN: 9781925726565

Number Bingo

Max was making Bingo Cards and Number Clues but he didn't have time to finish them. Please complete the Bingo Game for Max.

I made it so that every number clue covers two numbers on the Bingo Cards. So there are two 12s on the Bingo Cards to go with the number clue 3 × 4.

But I didn't have time to fill in all the cards or all the number clues.

Bingo Card

12	20	30
15	24	35

Bingo Card

	20	30
16		32

Bingo Card

12	18	
16		32

Bingo Card

	18	24
15		35

Number Clue	Number Clue	Number Clue	Number Clue
3 × 4	4 × 7	5 × 6	3 × 8
3 × 5	4 × 8	3 × 7	
5 × 4	4 × 4		

TARGETING MATHS HOMEWORK: YEAR 3 © PASCAL PRESS ISBN: 9781925726565

AC9M3N04, AC9M3A03

Multiplication and Division

You can use multiplication facts to help with division.

For example, if you know 6 × 5 = 30 and 5 × 6 = 30
then you also know 30 ÷ 5 = 6 and 30 ÷ 6 = 5.

Complete these multiplication and division fact families.

1. 4 × 6 = 24
   6 × 4 = 24
   24 ÷ 6 = ☐
   24 ÷ 4 = ☐

2. 5 × 7 = ☐
   7 × 5 = ☐
   ☐ ÷ 7 = ☐
   ☐ ÷ 5 = ☐

3. 3 × 9 = ☐
   9 × 3 = ☐
   ☐ ÷ 9 = ☐
   ☐ ÷ 3 = ☐

4. 6 × 7 = ☐
   7 × 6 = ☐
   ☐ ÷ 7 = ☐
   ☐ ÷ 6 = ☐

5. 7 × 3 = ☐
   3 × 7 = ☐
   ☐ ÷ 3 = ☐
   ☐ ÷ 7 = ☐

6. 8 × 3 = ☐
   ☐ × ☐ = ☐
   ☐ ÷ ☐ = ☐
   ☐ ÷ ☐ = ☐

7. 4 × 9 = ☐
   ☐ × ☐ = ☐
   ☐ ÷ ☐ = ☐
   ☐ ÷ ☐ = ☐

8. 7 × 8 = ☐
   ☐ × ☐ = ☐
   ☐ ÷ ☐ = ☐
   ☐ ÷ ☐ = ☐

9. 9 × 7 = ☐
   ☐ × ☐ = ☐
   ☐ ÷ ☐ = ☐
   ☐ ÷ ☐ = ☐

SCORE /9 0-4 5-7 8-9

Quick Check

Which multiplication facts help with these divisions? Write the missing numbers.

1. 24 ÷ 3 ☐ × ☐ = ☐
2. 16 ÷ 8 ☐ × ☐ = ☐
3. 20 ÷ 5 ☐ × ☐ = ☐
4. 18 ÷ 3 ☐ × ☐ = ☐
5. 21 ÷ 3 ☐ × ☐ = ☐
6. 90 ÷ 10 ☐ × ☐ = ☐
7. 27 ÷ 3 ☐ × ☐ = ☐
8. 16 ÷ 4 ☐ × ☐ = ☐
9. 25 ÷ 5 ☐ × ☐ = ☐
10. 36 ÷ 9 ☐ × ☐ = ☐

SCORE /10 0-4 5-8 9-10

TARGETING MATHS HOMEWORK: YEAR 3 © PASCAL PRESS ISBN: 9781925726565

Number & Algebra

UNIT 15

AC9M3N06

Division Word Problems

Write the division number sentences and then work out the answers.

1. Cook has twelve cherry tomatoes to put on three salad plates.
   How many tomatoes should she put on each plate?
   ☐ ÷ ☐ = ☐

2. Hugo bought five cartons of juice for twenty dollars.
   How much did each carton cost?
   ☐ ÷ ☐ = ☐

3. Ada counted sixteen wheels on the quad bikes.
   How many quad bikes were there?
   ☐ ÷ ☐ = ☐

4. The florist has fifteen roses to put into three vases.
   How many roses should he put in each vase?
   ☐ ÷ ☐ = ☐

5. Bindi made twenty-one $2 coins into three equal piles.
   What was the value of each pile?
   ☐ ÷ ☐ = ☐

SCORE /5 0-2 3-4 5

Statistics & Probability

AC9M3ST01

Bean Growth

Luis and Max are growing beans. They made graphs to record the height each week.

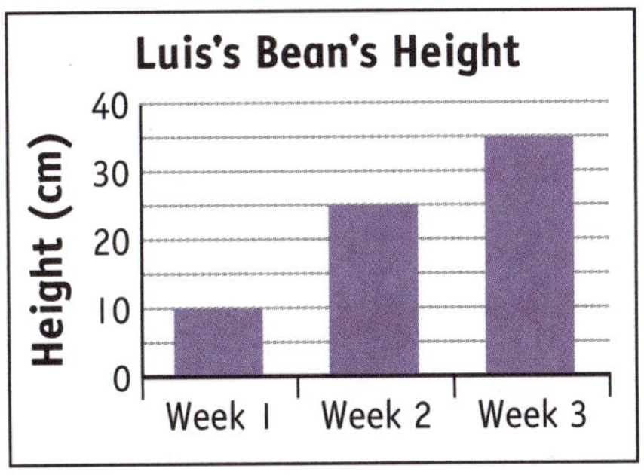

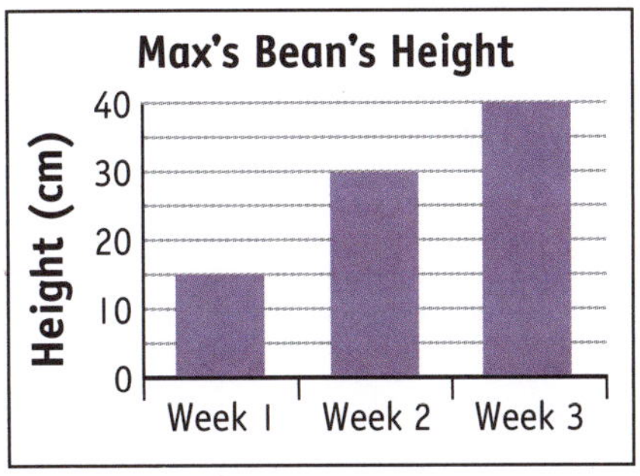

1. Whose bean was shorter in Week 1 and how tall was it?
   __________ bean was ____ cm tall.

2. Whose bean was taller in Week 1 and how tall was it?
   __________ bean was ____ cm tall.

3. How much did the beans grow between Week 1 and Week 2?
   Luis's bean grew ____ cm and Max's bean grew ____ cm.

4. How much did each bean grow altogether?
   Luis's bean grew ____ cm altogether and Max's bean grew ____ cm altogether.

5. Whose bean do you think will be taller by Week 4? Why?
   __________ bean will be taller because ______________________________.

SCORE /5 0-2 3-4 5

AC9M3M03

What Is the Time?

① How many hours are there between one midnight and the next midnight? ____ hours

② How many hours are there between midnight and noon? ____ hours

③ How many hours are there between noon and midnight? ____ hours

Write the letter of the arrow that points to a time:

④ in the morning. ____

⑤ in the evening. ____

⑥ in the afternoon. ____

TERM 2

Write the time.

⑦ A points to ____________ on the timeline.

⑧ B points to ____________ on the timeline.

⑨ C points to ____________ on the timeline.

⑩ On the timeline, mark 3 o'clock in the morning.

SCORE /10 (0-4) (5-8) (9-10)

NAPLAN*-style Questions

① Hugo has 24 balloons to put into 6 equal-sized bunches for his party. How many balloons should be in each bunch?

◯ 24 ◯ 12 ◯ 6 ◯ 4

② How many hours are between 3 o'clock in the morning and 3 o'clock in the afternoon?

[ ] hours

③ How much taller is the tomato plant than the bean plant?

[ ] cm

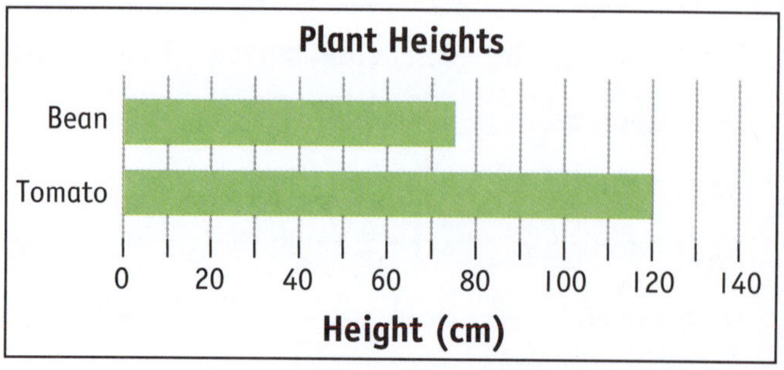

SCORE /3 (0-1) (2) (3)

TARGETING MATHS HOMEWORK: YEAR 3 © PASCAL PRESS ISBN: 9781925726565

Problem of the Week

Party Balloons

The party organiser has 72 balloons to use for table decorations.

Each table decoration must have the same number of balloons and he must use all the balloons.

This leaves him with a problem – there seem to be a lot of possibilities.

Help the party organiser work out the table decorations.

1. Can she make the decorations with three balloons in each?

   ⬭ Yes ⬭ No

   If so, how many decorations will there be?

   ____ decorations

2. Can she make the decorations with four balloons in each?

   ⬭ Yes ⬭ No

   If so, how many decorations will there be?

   ____ decorations

3. Can she make the decorations with five balloons in each?

   ⬭ Yes ⬭ No

   If so, how many decorations will there be?

   ____ decorations

4. What different numbers of balloons can be used in the table decorations with no balloons left over?

AC9M3N06

TERM 2

Work Smart Not Hard for Addition and Subtraction

Look at these three sums: **16 + 9 = __** **26 + 9 = __** **25 − 9 = __**

If you are not working smart, you will work out each one separately.

But if you are working smart, you will see that 16 + 9 = 25 and that 26 is 10 more than 16.

So 26 + 9 will be 10 more than 25, which is 35.

Now that you have done 16 + 9 = 25, 25 − 9 is a related subtraction: 25 − 9 = 16

Complete these sums.
Only work out the first addition each time and work smart to complete the next two sums.

1. 37 + 11 = __ 47 + 11 = __ 48 − 11 = __
2. 38 + 13 = __ 48 + 13 = __ 51 − 13 = __
3. 29 + 29 = __ 39 + 29 = __ 58 − 29 = __
4. 96 + 12 = __ 96 + 22 = __ 108 − 12 = __
5. 37 + 25 = __ 47 + 25 = __ 62 − 25 = __
6. 46 + 53 = __ 46 + 63 = __ 99 − 53 = __
7. 35 + 75 = __ 35 + 85 = __ 110 − 75 = __
8. 27 + 63 = __ 37 + 63 = __ 90 − 63 = __
9. 32 + 47 = __ 42 + 47 = __ 79 − 47 = __
10. 33 + 46 = __ 43 + 56 = __ 99 − 56 = __

SCORE /10 0-4 5-8 9-10

Quick Check

Complete each sum. Then draw lines to match the three sums that belong together.

1. 67 + 29 = __
2. 38 + 27 = __
3. 46 + 24 = __
4. 96 + 12 = __
5. 103 + 26 = __
6. 46 + 34 = __
7. 96 + 22 = __
8. 93 + 26 = __
9. 77 + 29 = __
10. 48 + 37 = __
11. 139 − 26 = __
12. 70 − 46 = __
13. 96 − 29 = __
14. 118 − 22 = __
15. 65 − 27 = __

SCORE /15 0-7 8-13 14-15

TARGETING MATHS HOMEWORK: YEAR 3 © PASCAL PRESS ISBN: 9781925726565

Estimation

Look quickly at the counters in the boxes and estimate how many counters are in each box.

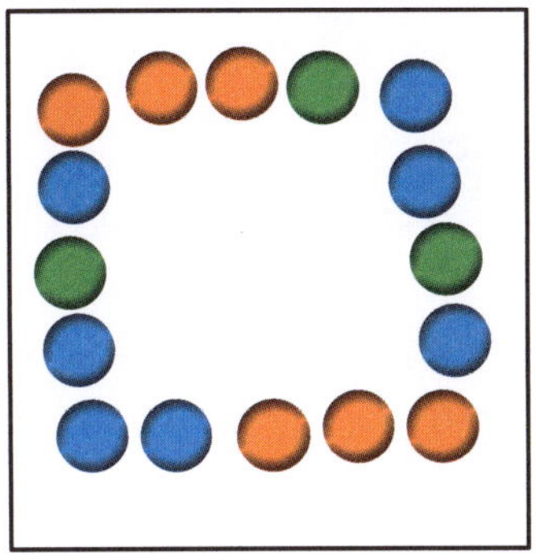

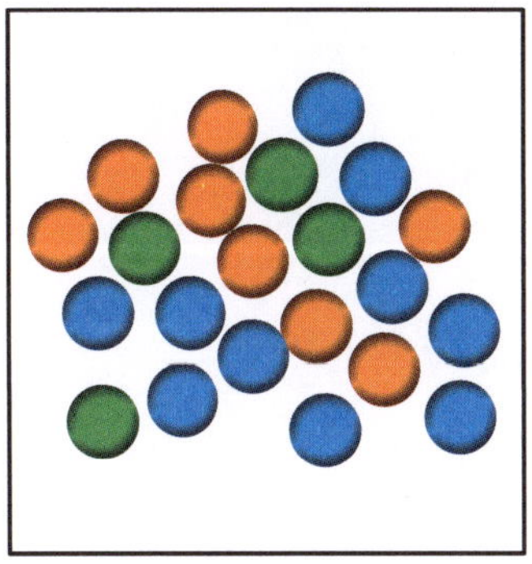

1. Estimate: ____
2. Estimate: ____
3. Estimate: ____

Now count to check how reasonable your estimates were.

4. Box 1 actual: ____
5. Box 2 actual: ____
6. Box 3 actual: ____
7. Which box was easiest to estimate? ________
8. Why was that? ______________________________

SCORE /8 0-3 4-6 7-8

Statistics & Probability

AC9M3P02

Rolling a Cube

Bindi stuck a coloured dot onto each face of a cube. She rolled the cube and made this tally of her results.

Bindi's Results	
Green	𝍸 |||
Blue	𝍸
Orange	||

1. How many times did Bindi roll the cube? _____
2. Which colour did she roll most often? ____________
3. How many green dots do you think she stuck on the cube? _____
4. How many blue dots do you think she stuck on the cube? _____
5. How many orange dots do you think she stuck on the cube? _____
6. What colour do you think she will roll next? ____________
7. Why? ______________________________

SCORE /7 0-3 4-5 6-7

AC9M3M02

Measuring Mass and Capacity

What is the mass?

① 

________ g

② 

________ g

③ 

________ g

How much liquid is in each jug?

④ 

Jug A

________ mL

⑤ 

Jug B

________ mL

⑥ 

Jug C

________ mL

⑦ How much does one apple weigh? __________

⑧ How much heavier are the two apples than the orange? __________

⑨ How much less liquid is in Jug A than in Jug C? __________

⑩ How many millilitres are in a quarter of a litre? __________

SCORE /10 0-4 5-8 9-10

NAPLAN*-style Questions

① One of these number sentences does not belong in the fact family.
Mark the number sentence that does not belong.

- ◯ 26 + 7 = ☐
- ◯ 37 + 15 = ☐
- ◯ 52 – 15 = ☐

② Luis weighed four cherries and then a pear.
About how much heavier is the pear than one cherry?

◯ 70 g ◯ 80 g ◯ 90 g ◯ 100 g

③ Look at the key. How many fish are there in the tanks altogether?

Key:  = 4 fish

☐

SCORE /3 0-1 2 3

TARGETING MATHS HOMEWORK: YEAR 3 © PASCAL PRESS ISBN: 9781925726565

AC9M3N05, AC9M3M03

Raindrops

One rainy day, Hugo filled a square of paper with raindrops.

1. Without counting, how many raindrops do you think Hugo drew on his square of paper?

   ____

2. How long do you think it took Hugo to draw the raindrops?

   ____

3. Check how many raindrops Hugo actually drew.

   ____ raindrops

Your turn

The grid is for you to draw raindrops in.

4. How many raindrops do you think you will need to fill the whole grid?

   ____

5. How long do you think it will take you to fill the whole grid?

   ____

Time yourself as you fill one box of the grid with raindrops.

6. How many raindrops are in the first box?

   ______

7. How long did it take to fill the first box?

   ____

8. Revise your estimates in question 4 and question 5.
9. Now complete the task and record the results.

   ____ raindrops. Time taken: ______________

TERM 2 REVIEW

Review: Unit 9 and Unit 10

AC9M3N02, AC9M3N05, AC9M3N06, AC9M3M06, AC9M3P01

(1) Estimate the number of blue dots.

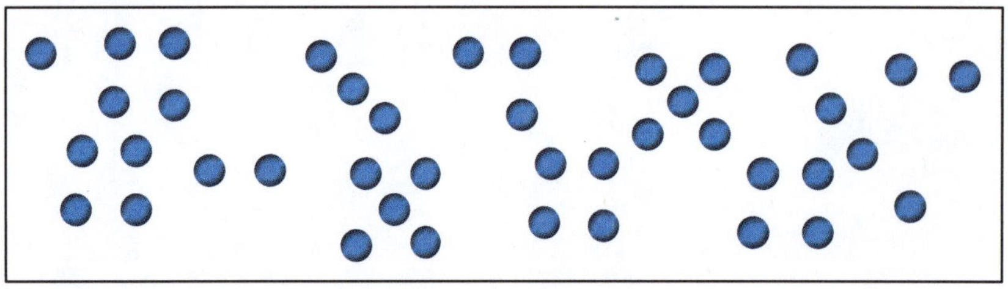

Estimate: _____ blue dots

(2) Count to check your estimate: _____ blue dots

Write a reasonable estimate for these additions.

(3) 86 – 29

Estimate: _____

(4) 76 + 38

Estimate: _____

(5) 6 × 12

Estimate: _____

(6) \$5·60 + \$7·50

Estimate: _______

(7) \$17·35 + \$10·35

Estimate: _______

(8) \$5·20 + \$6·50 + 35c

Estimate: _______

(9) \$12·70 + \$2 + 80c + \$3·30

Estimate: _______

(10) Four bread rolls at \$1·85 each

Estimate: _______

What fraction is shaded?

(11)  ―

(12) 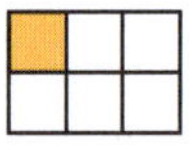 ―

(13) 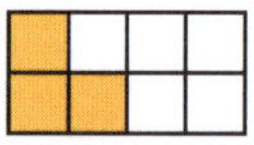 ―

(14) 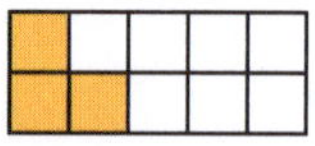 ―

What fraction is unshaded?

(15) 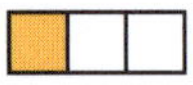 ―

(16) 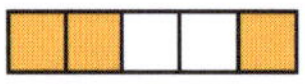 ―

(17) 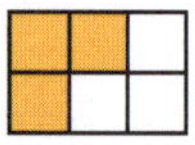 ―

(18) 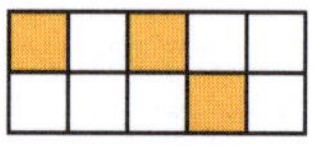 ―

(19) Which three coins make half of \$6·20?

_____, _____, _____

(20) Which five coins make \$1·75?

_____, _____, _____, _____, _____

(21) Is it likely to rain today?

◯ Yes ◯ No

(22) Explain your reasoning. ________________________________

SCORE /22 0-10 11-20 21-22

TARGETING MATHS HOMEWORK: YEAR 3 © PASCAL PRESS ISBN: 9781925726565

Review: Unit 11 and Unit 12

TERM 2 REVIEW

AC9M3N01, AC9M3N03, AC9M3M03, AC9M3M04, AC9M3ST01, AC9M3P02

Write these numbers using only digits.

(1) 15 hundreds and 15 ones ____________

(2) 10 thousands and 15 hundreds ____________

(3) How many tens make 2020? ____________

Use partitioning for these additions.

(4) 176 + 123		+		+		=		
(5) 2234 + 345		+		+		=		

Look at the chart.

(6) How many students were surveyed?

_____ students

(7) What might be a good title for this chart?

____

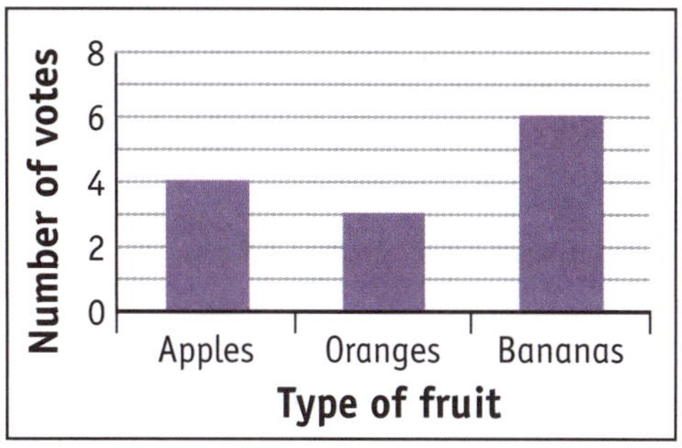

Show the times on the clocks.

Start Time: half past five	**Finish Time:** quarter to six	**Duration**
(8) 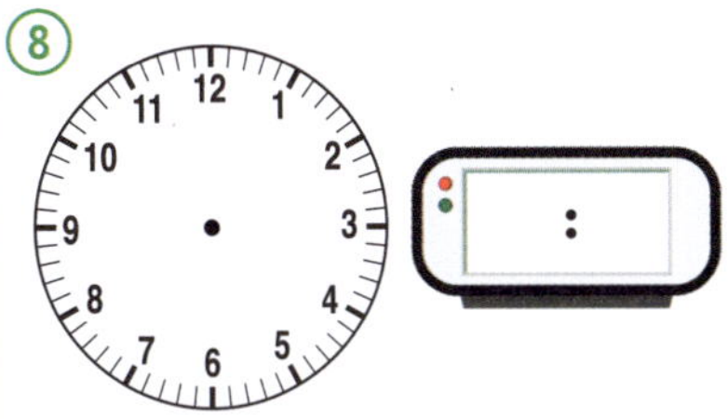	(9) 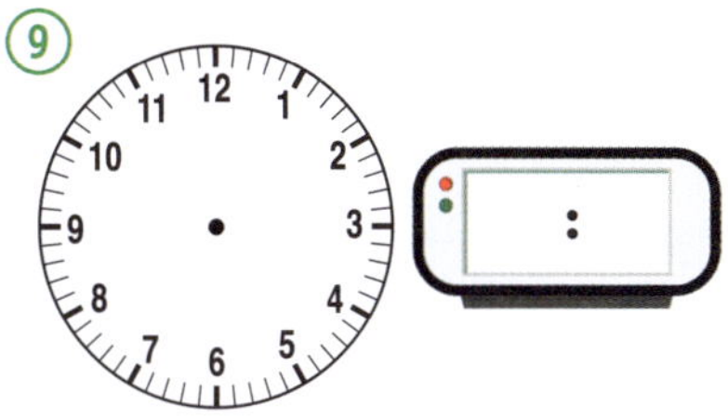	(10) How much time is there between half past five and quarter to six? ____________

Use partitioning for these subtractions.

(11) 266 − 142		+		+		=		
(12) 379 − 163		+		+		=		

(13) The chart is a tally of 20 spins of the spinner. Label the chart to show which colour belongs in each column.

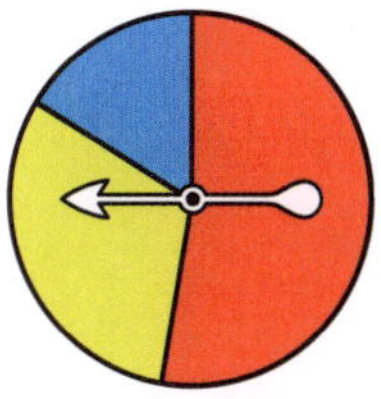

20 Spinner Spins		
𝍸 𝍸 |	𝍸 |	|||

SCORE /13 0-6 7-11 12-13

TERM 2

TERM 2 REVIEW

Review: Unit 13 and Unit 14

AC9M3N01, AC9M3N02, AC9M3N04, AC9M3M02

Are these statements true or false?

(1) 56 727 has 5 in the hundred thousands place.

◯ True ◯ False

(2) 370 698 has 0 in the thousands place.

◯ True ◯ False

(3) 168 439 has 6 in the ten thousands place.

◯ True ◯ False

(4) 976 999 is only 1 less than 977 000.

◯ True ◯ False

(5) Double five times. Start at 5.

☐

TERM 2

Answer these questions in grams (g), millilitres (mL) or centimetres (cm).

(6) How much is half a litre? ________

(7) How long is half a 30-centimetre ruler? ________

(8) How much is in a quarter of a kilogram? ________

(9) How much longer is two ruler lengths than half a metre? ________

Complete these multiplication facts.

(10) 4 × 9 = ☐

(11) 7 × 3 = ☐

(12) 5 × 11 = ☐

(13) 3 × 8 = ☐

Write the number sentences and the answers for these word problems.

(14) There were six bags of apples with seven apples in each bag.
How many apples altogether?

☐ × ☐ = ☐

(15) Each raffle ticket costs 30 cents.
How much are five raffle tickets?

☐ × ☐ = ☐

(16) Bindi folded a rectangular strip of paper twice to make this shape.
Then she unfolded the shape and marked the fold lines in red.
How many different sizes of angles did the fold lines make?

____ angles

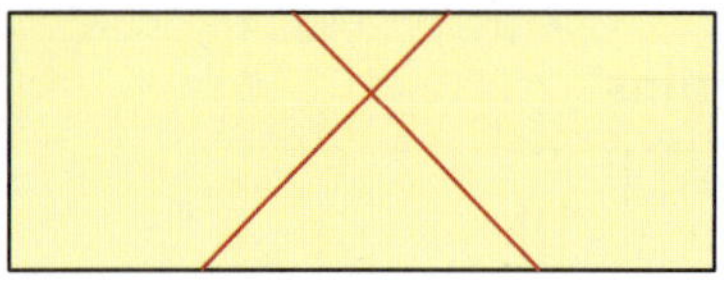

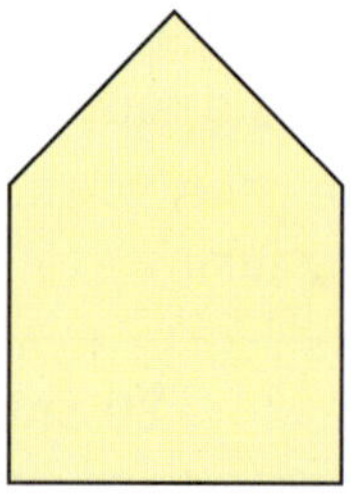

SCORE /16 0-7 8-14 15-16

TARGETING MATHS HOMEWORK: YEAR 3 © PASCAL PRESS ISBN: 9781925726565

Review: Unit 15 and Unit 16

AC9M3N04, AC9M3N06, AC9MA03, AC9M3M03, AC9M3ST01

Which multiplication facts will help with these divisions? Write the missing numbers.

(1) 12 ÷ 3 ☐ × 3 = 12

(2) 15 ÷ 5 ☐ × ☐ = ☐

(3) 24 ÷ 6 ☐ × ☐ = ☐

Write the division number sentences and then work out the answers.

(4) The art teacher has twenty-seven paintbrushes to put into three pots. How many paint brushes should she put in each pot?

_____________________ paintbrushes

(5) The florist has 24 daisies and needs to make three identical bouquets. How many daisies should he put into each bouquet?

_____________________ daisies

Look at the graph.

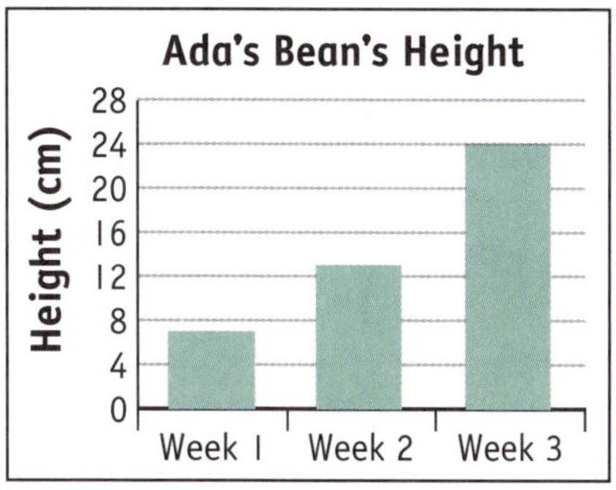

(6) How much did Ada's bean grow from Week 1 to Week 3?

_____ cm

(7) How much did Ada's bean grow from Week 1 to Week 2?

_____ cm

(8) How many hours are there between six in the morning and six in the evening?

_____ hours

Complete the related addition and subtraction facts.

(9) 37 + 13 = ☐ 37 + 23 = ☐ ☐ − 23 = 37

(10) 46 + 15 = ☐ 56 + 15 = ☐ ☐ − 15 = 46

How much does each piece of fruit weigh?

(11)

Each kiwi fruit weighs ________.

(12)

Each nectarine weighs ________.

SCORE /12 0-5 6-10 11-12

TARGETING MATHS HOMEWORK: YEAR 3 © PASCAL PRESS ISBN: 9781925726565

Pattern Rules

Follow the pattern rules. Show the jumps on the pattern sequence lines.

① Add 3, subtract 2.

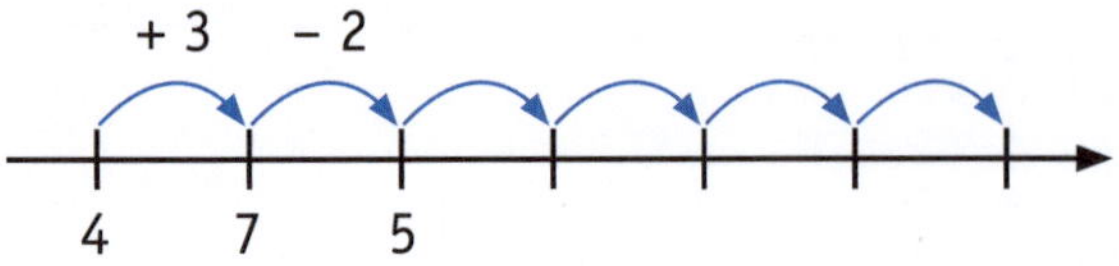

② Add 6.

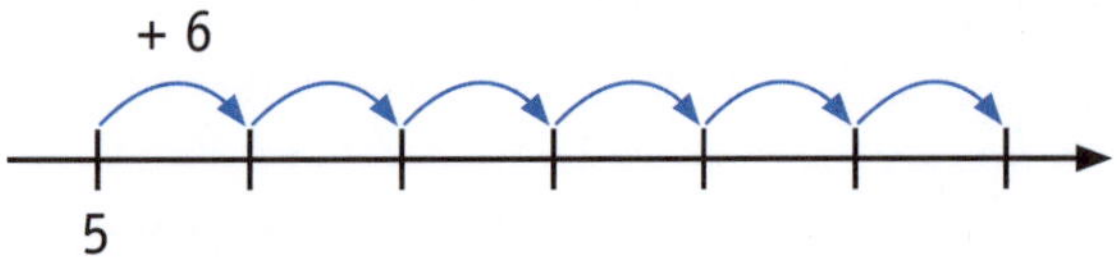

③ Double, subtract 2.

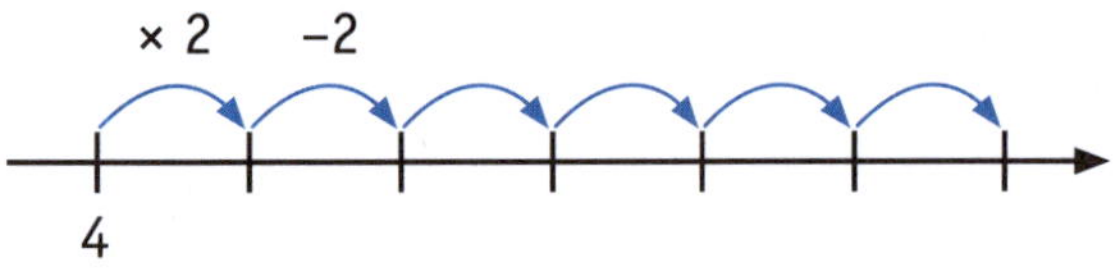

④ Add 3, double.

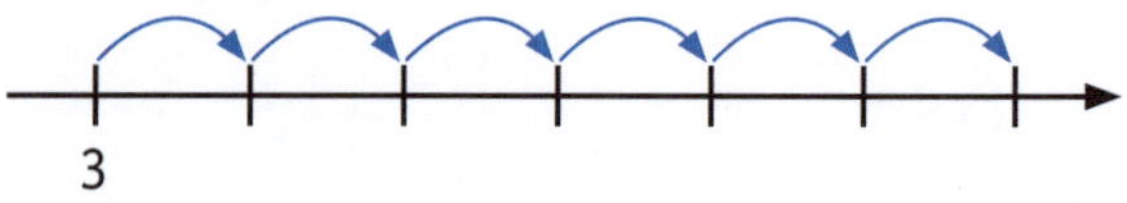

⑤ Add 4, subtract 1.

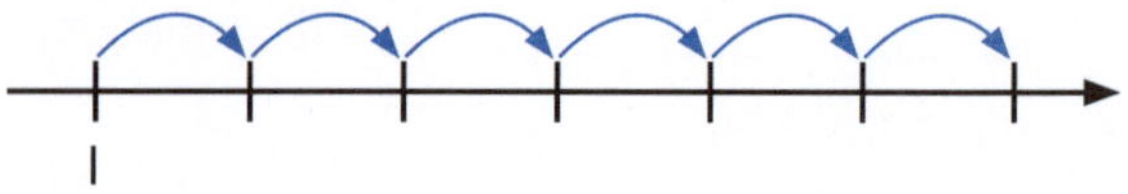

Continue these patterns on the pattern sequence lines. Write the rules.

⑥ 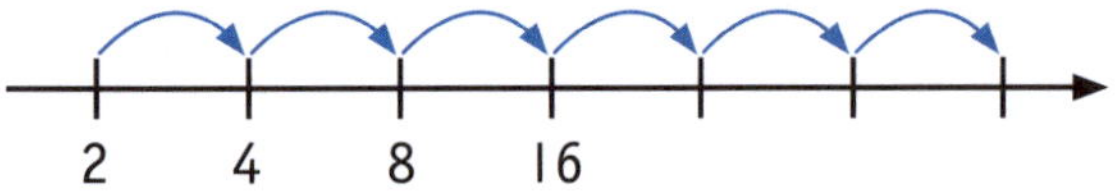

Rule: ______________________

⑦ 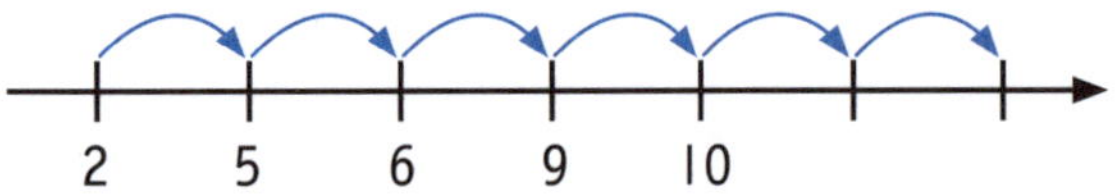

Rule: ______________________

⑧ 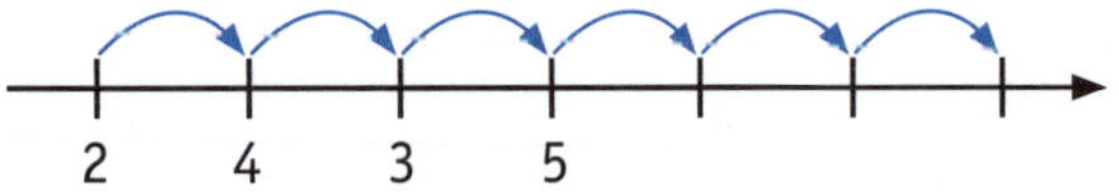

Rule: ______________________

SCORE /8 0-3 4-6 7-8

Quick Check

Continue these pattern sequences.

① 3, 6, 7, 10, 11, _____, _____, _____

② 13, 24, 35, _____, _____, _____

③ 26, 25, 21, 20, 16, _____, _____, _____

④ 3, 6, 5, 8, 7, _____, _____, _____

⑤ 12, 13, 15, 18, 22, _____, _____, _____

⑥ 12, 10, 20, _____ 28, _____, _____

⑦ 6, 11, 16, _____, _____, _____

⑧ 94, 95, 98, 99, _____, 103, _____, _____, _____

⑨ 93, 96, 95, 98, 97, _____, _____, _____

⑩ 93, 96, 99, _____, _____, _____

SCORE /10 0-4 5-8 9-10

TARGETING MATHS HOMEWORK: YEAR 3 © PASCAL PRESS ISBN: 9781925726565

Number & Algebra

AC9M3A01

Unknown Values

In this problem, the answer is unknown: **63 + 6 = ?**

Complete these **answer unknown** problems.

1. 63 + 6 = ☐
2. 48 + 8 = ☐
3. 54 + 67 = ☐
4. 69 − 6 = ☐
5. 56 − 48 = ☐
6. 121 − 67 = ☐

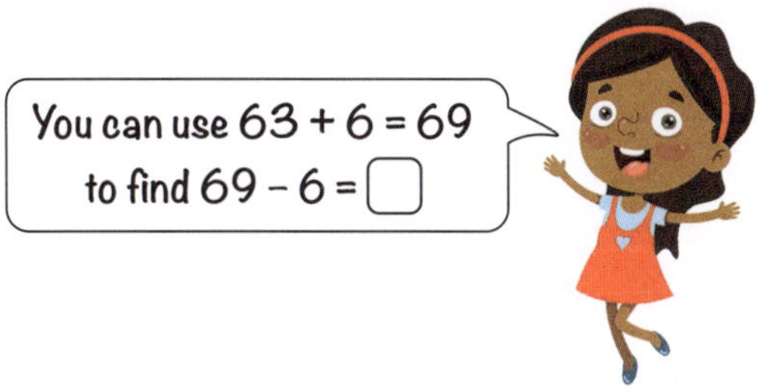

In this problem, the start number is unknown: **? + 6 = 69**

Complete these **start unknown** problems.

7. ☐ + 74 = 96
8. ☐ + 48 = 57
9. ☐ + 64 = 71
10. 22 + ☐ = 96
11. 9 + ☐ = 57
12. 7 + ☐ = 71

SCORE /12 0-5 6-10 11-12

Statistics & Probability

AC9M3ST01, AC9M3M06

Making a Graph

Max has been collecting coins and he decided to sort them. He made this chart to record his results.

How many coins and how much money does Max have altogether?

Use the table to work out the answers.

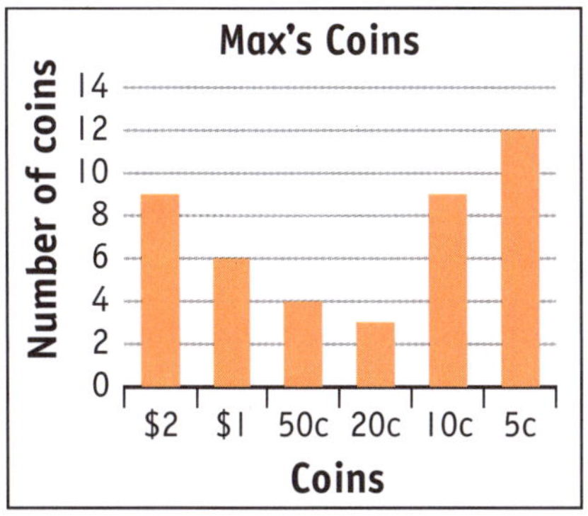

	Coin	Number	Value
1	$2		
2	$1		
3	50c		
4	20c		
5	10c		
6	5c		
7	**Total**		

8. Max has ____ coins with a total value of __________.

SCORE /8 0-3 4-6 7-8

TARGETING MATHS HOMEWORK: YEAR 3 © PASCAL PRESS ISBN: 9781925726565

Measurement & Space

AC9M3M02

Estimating and Measuring Length

Use these lines to answer the following questions.

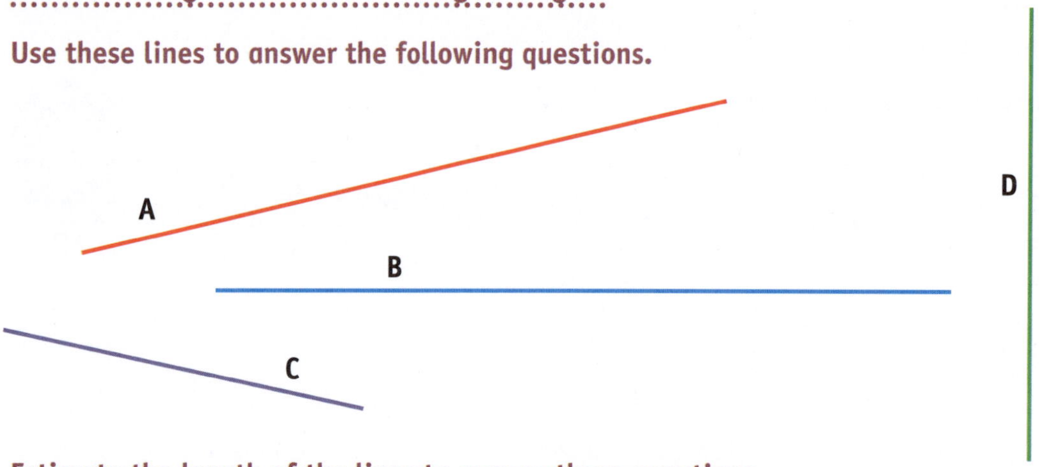

Estimate the length of the lines to answer these questions.

1. The longest line is Line ____.
2. The shortest line is Line ____.
3. Line ____ is twice as long as Line ____.
4. Line ____ is 3 cm longer than Line ____.

Use a ruler to measure the lengths of the lines to answer these questions.

5. The longest line is ____ cm.
6. The shortest line is ____ cm.
7. Line ____ is half as long as Line ____.
8. Line A and Line B together are ____ cm longer that Line C and Line D together.

TERM 3

SCORE /8 0-3 4-6 7-8

NAPLAN*-style Questions

1. What number should be in the yellow square? ____

12		16		
22				?

2. What is the unknown number in this equation? ☐ + 48 = 64

◯ 48 ◯ 112 ◯ 16 ◯ 12

3. Line A is 3 cm long.

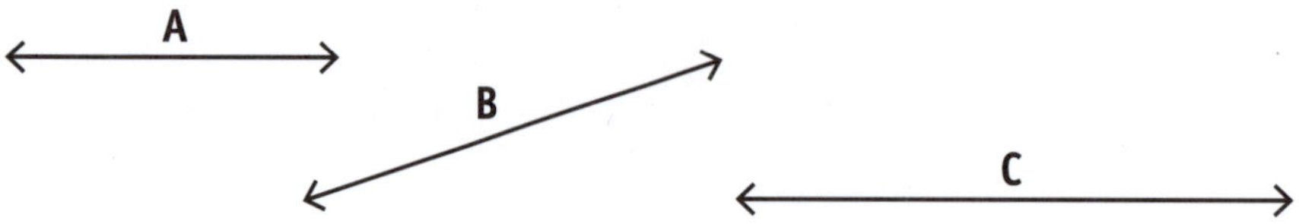

What is the combined length of Lines A, B and C? ____ cm

SCORE /3 0-1 2 3

TARGETING MATHS HOMEWORK: YEAR 3 © PASCAL PRESS ISBN: 9781925726565

Problem of the Week

UNIT 17

AC9M3M02

How Long Are These Lines?

Hugo drew the red line, Ada drew the blue line and Bindi drew the purple line. Before using a ruler to measure, estimate answers to these questions.

1. Who drew the longest line? ____________
2. Who drew the shortest line? ____________
3. About how long do you think the purple line is? ____________

4. Use your ruler to find the actual lengths of the lines.

   Red line = _____ cm Blue line = _____ cm Purple line = _____ cm

5. Were you surprised? ◯ Yes ◯ No

   Why was that? ____

TERM 3

Fractions

Luis cut up four circles of yellow paper to make a fraction puzzle.

The fractions cut from a circle were all the same.

Then someone opened the classroom door and some of his pieces blew away.

The picture shows the fraction pieces that were left.

What size fraction pieces did Luis make?

① ________ ② ________ ③ ________ ④ ________

Remember that we can count in fractions: $\frac{1}{5}$, $\frac{2}{5}$, $\frac{3}{5}$, $\frac{4}{5}$, $\frac{5}{5}$ or 1 whole

What size are the fraction pieces that are missing?

⑤ $\frac{\square}{\square}$ ⑥ $\frac{\square}{\square}$ ⑦ $\frac{\square}{\square}$

How many fraction pieces were blown away?

⑧ ____ pieces are missing.

SCORE /8 0-3 4-6 7-8

Quick Check

The shaded part of each diagram makes a fraction of the whole. Name the shaded fraction and that fraction that remains unshaded.

		Shaded Fraction	Unshaded Fraction
①	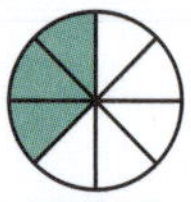		
②	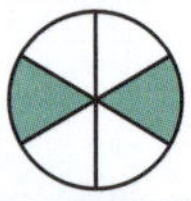		
③	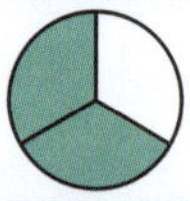		
④			
⑤	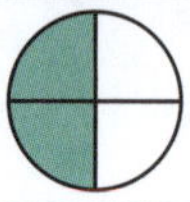		

		Shaded Fraction	Unshaded Fraction
⑥	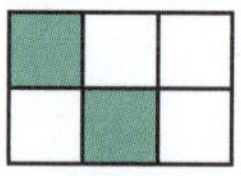		
⑦	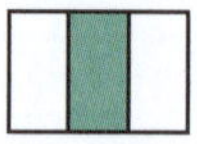		
⑧	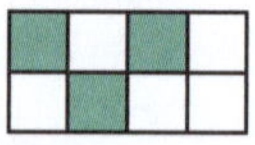		
⑨	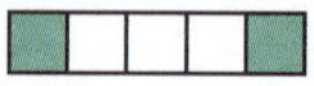		
⑩		Has this circle been divided into thirds? ◯ Yes ◯ No	

SCORE /10 0-4 5-8 9-10

TARGETING MATHS HOMEWORK: YEAR 3 © PASCAL PRESS ISBN: 9781925726565

Number & Algebra

AC9M3A01

Unknown Values in Addition

In this problem, the change is unknown: 63 + ? = 69

We know the start number but not what was added to it to get the answer, 69.
You can write this problem as: **What should be added to 63 to get to 69?**

Complete these change unknown problems.

1. 74 + ☐ = 79
2. 53 + ☐ = 61
3. 47 + ☐ = 57
4. 36 + ☐ = 62
5. 26 + ☐ = 43
6. 37 + ☐ = 59
7. 98 + ☐ = 101
8. 26 + ☐ = 126
9. 37 + ☐ = 69
10. 86 + ☐ = 116
11. 38 + ☐ = 44
12. 76 + ☐ = 85
13. 27 + ☐ = 37
14. 38 + ☐ = 49
15. 53 + ☐ = 65
16. 27 + ☐ = 47
17. 55 + ☐ = 66
18. 88 + ☐ = 108
19. 20 + ☐ = 106
20. 193 + ☐ = 210

SCORE /20 0-9 10-18 19-20

Statistics & Probability

AC9M3ST01

Data Collection

The graph shows how many students under 8 years and how many students 8 years and over do activities after school.

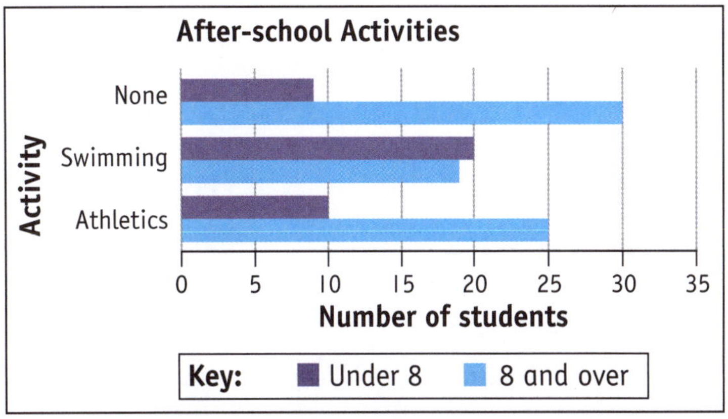

1. How many students do none of the after-school activities? ______
2. Which after-school activity is most popular with under 8s? ______
3. How many more 8 and overs than under 8s do athletics? ______
4. How many students were surveyed? ______
5. How many more 8 and overs than under 8s do after-school activities? ______

SCORE /5 0-2 3-4 5

AC9M3M06

Money Amounts

Kiku found four ways of making 20 cents. Notice how she changed only one coin on each row.

Method 1

Method 2

Method 3

Method 4

Use Kiku's method to find five different ways of making 30 cents.

1. Method 1: ______________________
2. Method 2: ______________________
3. Method 3: ______________________
4. Method 4: ______________________
5. Method 5: ______________________
6. There is one more way to make 30 cents. Can you work out what it is?

____

SCORE /6 0-2 3-4 5-6

TERM 3

NAPLAN*-style Questions

1. Hugo was counting down from 27 in 2s. He counted 27, 25, 23, 21. What number should he count next?

   ◯ 20 ◯ 23 ◯ 19 ◯ 18

2. At training last night there were 14 boys and 5 more girls than boys. Which tally table matches this data?

◯

Boys	𝍸 𝍸 ||||
Girls	||||

◯

Boys	𝍸 𝍸 ||||
Girls	𝍸 𝍸 𝍸 ||||

◯

Girls	𝍸 𝍸 ||||
Boys	𝍸 𝍸 𝍸 ||||

3. Ada had these coins. Then she spent $3·65.

How much money did Ada have left?

◯ 5c ◯ 10c ◯ 15c ◯ 20c

SCORE /3 0-1 2 3

TARGETING MATHS HOMEWORK: YEAR 3 © PASCAL PRESS ISBN: 9781925726565

AC9M3N06, AC9M3M06

Who Had Which Coins?

Bindi, Luis and Max each had $3.
These are the coins that they had between them.
Who had which coins?

Bindi's Clue

I don't have any gold coins,
but I do have ten silver coins.

Luis's Clue

I have two gold coins but
I only have five silver coins.

Max's Clue

I have one gold coin
and six silver coins.

Ada has a hint for you!

Match the picture with real coins or
make your own so that you can guess,
check and improve to find the solution.

AC9M3A03

Multiplication

Complete these multiplication facts.

1. 3 × 3 = ☐
2. 3 × 4 = ☐
3. 3 × 5 = ☐
4. 3 × 6 = ☐
5. What do you notice about the pattern in the answers?

____

Use the pattern to help you work out these multiplications.

6. 3 × 23 = ☐
7. 3 × 24 = ☐
8. 3 × 25 = ☐
9. 3 × 26 = ☐

Complete these multiplications by noticing and using their patterns.

10. 4 × 3 = ☐
11. 4 × 4 = ☐
12. 4 × 5 = ☐
13. 4 × 6 = ☐
14. 4 × 33 = ☐
15. 4 × 34 = ☐
16. 4 × 35 = ☐
17. 4 × 36 = ☐
18. 5 × 3 = ☐
19. 5 × 5 = ☐
20. 5 × 7 = ☐
21. 5 × 9 = ☐
22. 5 × 23 = ☐
23. 5 × 25 = ☐
24. 5 × 27 = ☐
25. 5 × 29 = ☐
26. 10 × 3 = ☐
27. 10 × 6 = ☐
28. 10 × 9 = ☐
29. 10 × 12 = ☐
30. 10 × 13 = ☐
31. 10 × 16 = ☐
32. 10 × 19 = ☐
33. 10 × 22 = ☐

SCORE /33 0-16 17-31 32-33

Quick Check

Work smart to answer these questions.

1. 6 × 5 = ☐
2. 5 × 6 = ☐
3. 3 × 6 = ☐
4. 6 × 3 = ☐
5. 10 × 3 = ☐
6. 3 × 10 = ☐
7. 6 × 4 = ☐
8. 4 × 6 = ☐
9. 10 × 7 = ☐
10. 11 × 4 = ☐
11. 7 × 5 = ☐
12. 6 × 6 = ☐
13. 6 × 7 = ☐
14. 7 × 3 = ☐
15. 10 × 13 = ☐
16. 14 × 10 = ☐
17. 7 × 4 = ☐
18. 5 × 4 = ☐
19. 11 × 3 = ☐
20. 12 × 4 = ☐

SCORE /20 0-9 10-18 19-20

TERM 3

TARGETING MATHS HOMEWORK: YEAR 3 © PASCAL PRESS ISBN: 9781925726565

Number & Algebra

AC9M3A03

Multiplication and Division

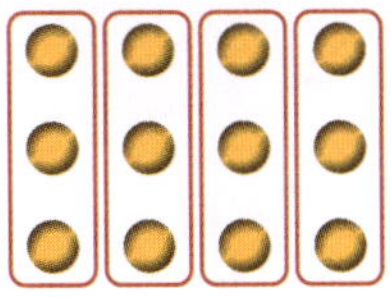

The columns in this array of counters are looped to show that 4 groups of 3 makes 12.

4 × 3 = 12

The loops also show that 12 shared into 4 groups makes 4 groups of 3.

12 ÷ 4 = 3

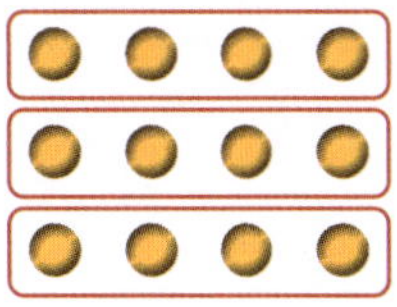

This is the same array of counters but this time the loops show 3 groups of 4.

3 × 4 = 12

12 ÷ 3 = 4

Show two ways of looping these groups. Write the related multiplication and division facts.

①

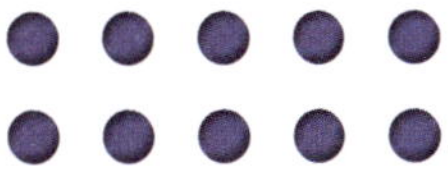

___ × ___ = ___

___ ÷ ___ = ___

②

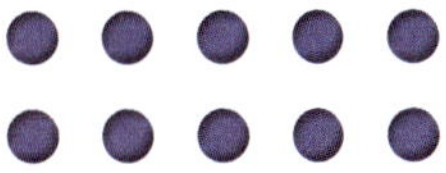

___ × ___ = ___

___ ÷ ___ = ___

③

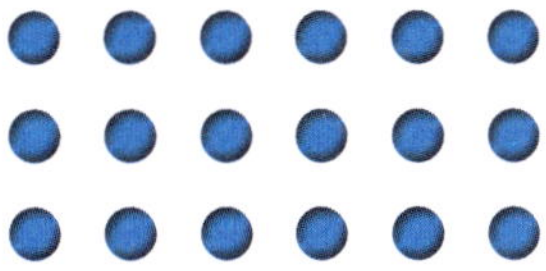

___ × ___ = ___

___ ÷ ___ = ___

④

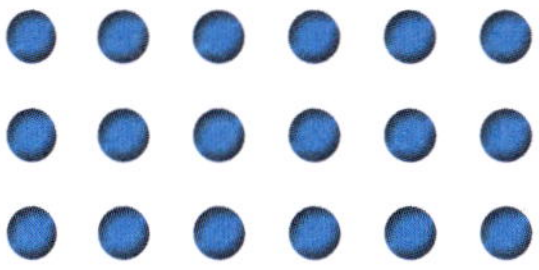

___ × ___ = ___

___ ÷ ___ = ___

SCORE /12 0-5 6-10 11-12

Measurement & Space

AC9M3M03

Calendar Months

Use the calendar to answer these questions.

① What day of the week is 14th June?

____

② What is the date of the first Monday in June?

____

③ What will the date be 10 days after the last Saturday in June?

____

④ What was the month before June?

____

⑤ In which season is June? ________________

⑥ Which other months have the same number of days as June?

____

June						
Sun.	**Mon.**	**Tue.**	**Wed.**	**Thu.**	**Fri.**	**Sat.**
28	29	30	31	1	2	3
4	5	6	7	8	9	10
11	12	13	14	15	16	17
18	19	20	21	22	23	24
25	26	27	28	29	30	1
2	3	4	5	6	7	8

SCORE /6 0-2 3-4 5-6

AC9M3P02

Any Chance?

Bindi has put 10 coloured counters into a bag: 2 red, 2 blue, 2 yellow, 2 green and 2 black. She will shake the bag and, without looking, pull out one counter.

1. Colour the counters below to match the 10 counters in Bindi's bag.

◯ ◯ ◯ ◯ ◯ ◯ ◯ ◯ ◯ ◯

2. What is the chance of Bindi pulling out a black counter?
   ◯ Impossible ◯ Unlikely ◯ Likely ◯ Very likely ◯ Certain

3. What is the chance of Bindi pulling out a red counter?
   ◯ Impossible ◯ Unlikely ◯ Likely ◯ Very likely ◯ Certain

4. What is the chance of Bindi not pulling out a black or a red counter?
   ◯ Impossible ◯ Unlikely ◯ Likely ◯ Very likely ◯ Certain

5. What is the chance of Bindi pulling out a white counter?
   ◯ Impossible ◯ Unlikely ◯ Likely ◯ Very likely ◯ Certain

6. What is the chance of Bindi pulling out a a counter that was not purple?
   ◯ Impossible ◯ Unlikely ◯ Likely ◯ Very likely ◯ Certain

7. Write your own chance question about Bindi's counters. Then write the answer.

Question: ______________________________

____

Answer: ______________

SCORE /5 0-2 3-4 5

TERM 3

NAPLAN*-style Questions

1. Which of these could you use to work out how many legs are on six ants?
   ◯ 8 ◯ 8 × 8
   ◯ 6 + 6 ◯ 6 × 6

2. Which two number sentences match this looped array?
   ◯ 5 × 3 = 15
   ◯ 15 ÷ 3 = 5
   ◯ 3 × 5 = 15
   ◯ 15 ÷ 5 = 3

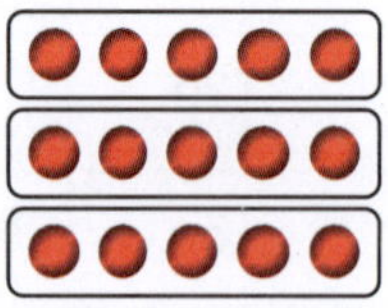

3. Bindi's birthday is marked on the calendar. Luis's birthday is 5 days later. On what day of the week is Luis's birthday?

____

June

Sun.	Mon.	Tue.	Wed.	Thu.	Fri.	Sat.
28	29	30	31	1	2	3
4	5	6	7	8	9	10
11	12	13	14	15	16	17
18	19	20	21	22	23	24
25	26	27	28	29	30	1
2	3	4	5	6	7	8

SCORE /3 0-1 2 3

TARGETING MATHS HOMEWORK: YEAR 3 © PASCAL PRESS ISBN: 9781925726565

Problem of the Week

AC9M3A03, AC9M3N06

The Chef Challenge

Team Blue and Team Red must make as many identical pizzas as they can using all of their ingredients.

Team Blue's ingredients:
- 36 olives
- 12 mozzarella balls
- 18 salami slices
- 15 pineapple chunks

Team Red's ingredients:
- 24 mini-meatballs
- 6 anchovies cut in halves
- 16 tomato slices
- 32 olives

Both teams have as much flour, milk and tomato paste as they need to make their pizzas.

(1) How many identical pizzas can Team Blue make?

____ pizzas

(2) Show what one of their pizzas would look like and list the ingredients.

Ingredients

____
____
____
____
____
____

(3) How many identical pizzas can Team Red make?

____ pizzas

(4) Show what one of their pizzas would look like and list the ingredients.

Ingredients

____
____
____
____
____
____

AC9M3N01

Counting Play Money

Four students were playing a game with play money.

| $100 | $50 | $10 | $5 | $1 |

These are the notes that each player had at the end of the first game.

Max: I have three $100 notes, two $50 notes, a $10 note and a $1 note.

Kiku: I have eleven $10 notes, a $5 note and six $1 notes.

Hugo: I have a $100 note and four $50 notes.

Ada: I have twenty-two $10 notes, three $100 notes and six $1 notes.

How much money did each player have?

1. Max: _______
2. Kiku: _______
3. Hugo: _______
4. Ada: _______

Now answer these questions.

5. ________ had the most money.
6. ________ had the least money.
7. How much more money did Ada have than Kiku? _______
8. How much less money did Hugo have than Max? _______
9. How much more money did Max have than Kiku? _______
10. How much money did the four players have altogether? _______

SCORE /10 0-4 5-8 9-10

TERM 3

Quick Check

How much are the notes worth?

1. 3 hundreds, 11 tens, 6 ones
   $__________
2. 11 hundreds, 6 tens, 5 ones
   $__________
3. 9 hundreds, 11 tens, 15 ones
   $__________
4. 12 hundreds, 12 tens, 12 ones
   $__________
5. 2 thousands, 9 hundreds, 9 tens, 11 ones
   $__________
6. 23 hundreds, 6 tens, 5 ones
   $__________
7. 10 thousands, 6 hundreds, 5 tens, 6 ones
   $__________
8. 9 hundreds, 12 tens, 15 ones
   $__________
9. 20 thousands, 16 hundreds, 16 tens
   $__________
10. 4 thousands, 6 hundreds, 30 tens, 6 ones
    $__________

SCORE /10 0-4 5-8 9-10

TARGETING MATHS HOMEWORK: YEAR 3 © PASCAL PRESS ISBN: 9781925726565

Making Friendly Numbers for Addition

You can use friendly number pairs like 6 and 4 or 8 and 2 to make additions easier.

$76^{+4} + 64^{-4}$
becomes
$80 + 60 = 140$

Choose the number closer to a ten: 76
Add to make 10: 76 + 4 = 80
Subtract the 4 from the second number to balance. 64 – 4 = 60
80 + 60 is easier to work out than 76 + 64.

Here is another example:

$178^{+2} + 122^{-2}$
becomes
$180 + 120 = 300$

Use friendly numbers to complete these additions.

1. $27^{+3} + 53^{-3}$
   ☐ + ☐ = ☐
2. $28^{+2} + 42^{-2}$
   ☐ + ☐ = ☐
3. $46^{+\_\_} + 84^{-\_\_}$
   ☐ + ☐ = ☐
4. 38 ___ + 46 ___
   ☐ + ☐ = ☐
5. 29 ___ + 64 ___
   ☐ + ☐ = ☐
6. 77 ___ + 25 ___
   ☐ + ☐ = ☐
7. 38 ___ + 83 ___
   ☐ + ☐ = ☐
8. 23 ___ + 89 ___
   ☐ + ☐ = ☐
9. 85 ___ + 26 ___
   ☐ + ☐ = ☐
10. 98 ___ + 87 ___
   ☐ + ☐ = ☐

SCORE /10 0-4 5-8 9-10

Statistics & Probability

AC9M3P01

Counters in a Bag

Luis put six counters into a bag.

There were two counters with the number 1, two with the number 2 and two with the number 3.

Without looking, Luis took two counters out of the bag. He added the numbers on them and then put the counters back in the bag and gave it a shake.

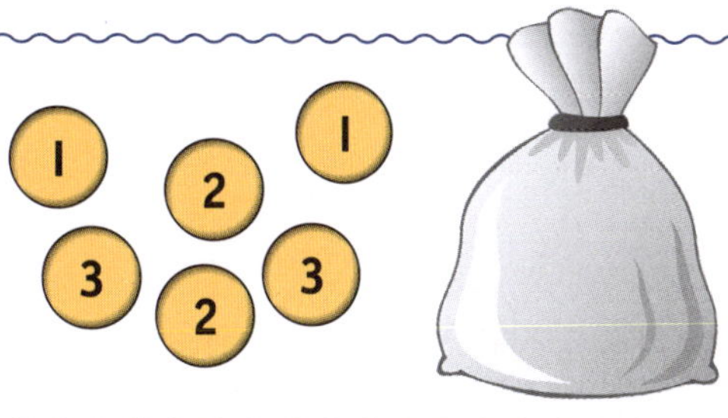

1. What are the possible totals that Luis could make with the two counters he took out?

   ____

2. Is one of the totals more likely to happen than any of the others?

   ◯ Yes ◯ No

3. If so, what is the most likely total? ____

SCORE /3 0-1 2 3

AC9M3SP01

Finding Shapes in the Environment

Look around your environment and see if you can find five objects that are cylindrical or almost cylindrical.

Then see if you can find five objects that are almost spherical. Here are a couple of ideas to get you started.

Five objects like cylinders

① ____________________

② ____________________

③ ____________________

④ ____________________

⑤ ____________________

Five objects like spheres

⑥ ____________________

⑦ ____________________

⑧ ____________________

⑨ ____________________

⑩ ____________________

SCORE /10 (0-4) (5-8) (9-10)

TERM 3

NAPLAN*-style Questions

① Hugo entered this number on his calculator:
He then pressed some more keys to change the display to 773.
Which key sequence did he press?

- ◯ [+] [1] [=]
- ◯ [+] [1] [0] [=]
- ◯ [+] [1] [0] [0] [=]

② Ada added two of these cards.
Which of these totals is it impossible for her to make?

[2] [3] [4]

◯ 5   ◯ 6   ◯ 7   ◯ 8

③ Which object are all of these things most like?

- ◯ Cube
- ◯ Rectangular prism
- ◯ Cylinder
- ◯ Pyramid

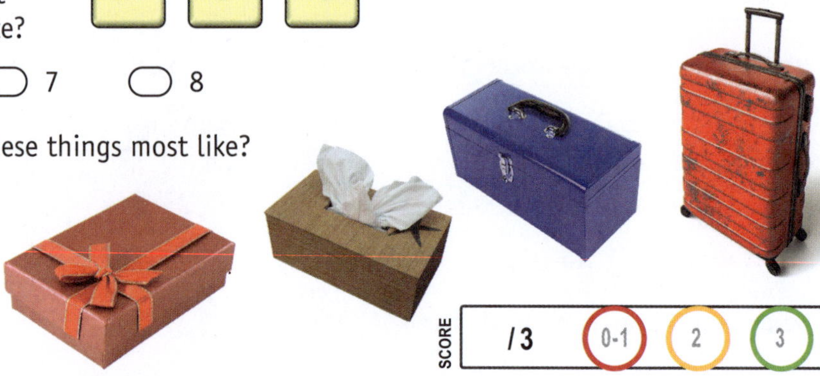

SCORE /3 (0-1) (2) (3)

TARGETING MATHS HOMEWORK: YEAR 3 © PASCAL PRESS ISBN: 9781925726565

Problem of the Week

AC9M3N01

Closest to $5000

$1000	$100	$50	$10	$5

To play **Closest to $5000**, play money notes are shuffled and put on the table facedown. On their turn, players take three notes from the pile and set them out in place value order.

At the end of three rounds, the players had these notes:

Bindi: 2 × $1000, 1 × $50, 6 × $10

Luis: 2 × $1000, 5 × $100, 1 × $50, 1 × $5

Max: 3 × $1000, 1 × $100, 1 × $10, 4 × $5

Kiku: 3 × $1000, 5 × $100, 1 × $10

1. How much money did each player have so far?

   Bindi: ______________ Max: ______________

   Luis: ______________ Kiku: ______________

2. Which player is closest to $5000 at the end of Round 3? ______________

Round 4

3. Which three notes does each player need to get as close to $5000 as possible? Complete the table to help you work it out.

Player	Current Score	Notes Needed	Final Score
Bindi			
Luis			
Max			
Kiku			

Imagine that each player gets the notes they need.

4. Which player will be closest to $5000 after Round 4? ______________
5. Which player will be furthest away from $5000? ______________

TERM 3

Estimation

If you take a quick look at these tens frames, you can estimate that there are more than 20 counters and a little less than 30 counters.

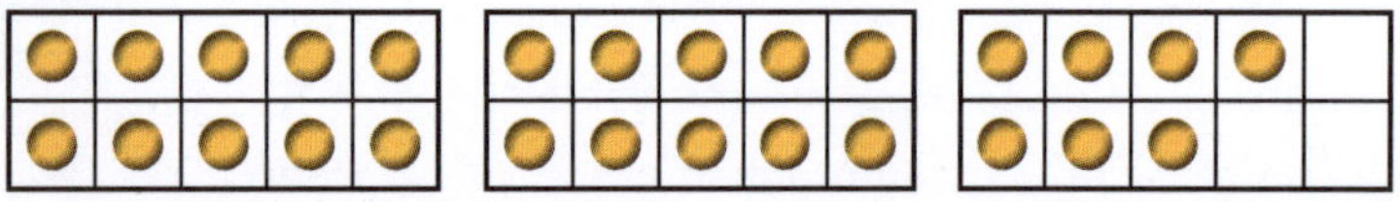

So a reasonable estimate would be **about 30** rather than **about 20**.

Make quick estimates for these tens frames.

1. Estimate: ____

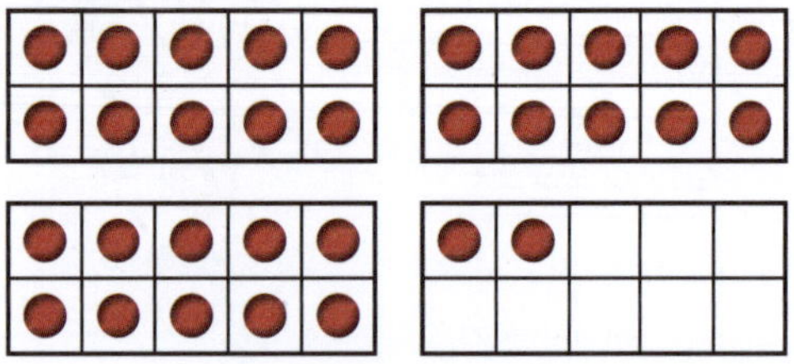

2. Estimate: ____

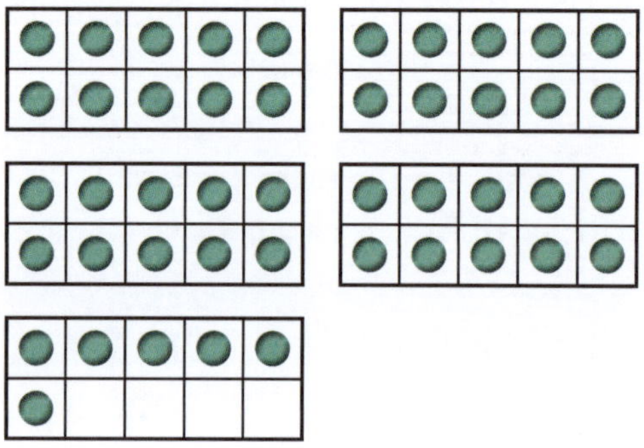

If you were estimating 28 + 24, you could change 28 to 30, the nearest friendly number, and think 30 + 24 = 54. This gives a reasonable estimate of 28 + 24.

Show how to estimate in these examples. The first one is started for you.

3. For 29 + 33, use [30] + [33] = [ ]
4. For 36 + 28, use [ ] + [ ] = [ ]
5. For 33 + 69, use [ ] + [ ] = [ ]
6. For 63 + 38, use [ ] + [ ] = [ ]
7. For 57 + 24, use [ ] + [ ] = [ ]
8. For 47 + 24, use [ ] + [ ] = [ ]
9. For 76 + 29, use [ ] + [ ] = [ ]
10. For 99 + 13, use [ ] + [ ] = [ ]

SCORE /10 0-4 5-8 9-10

Quick Check

Write estimates for these additions.

1. 28 + 9
   Estimate: ____
2. 47 + 28
   Estimate: ____
3. 34 + 57
   Estimate: ____
4. 17 + 34
   Estimate: ____
5. 58 + 36
   Estimate: ____
6. 73 + 28
   Estimate: ____
7. 84 + 18
   Estimate: ____
8. 23 + 49
   Estimate: ____
9. 18 + 37
   Estimate: ____
10. 87 + 28
    Estimate: ____

SCORE /10 0-4 5-8 9-10

TARGETING MATHS HOMEWORK: YEAR 3 © PASCAL PRESS ISBN: 9781925726565

Number & Algebra

AC9M3A02

Addition Using Pairs That Make 10

Pairs that make 10 can help with pairs that make 100 or even 1000.
If you know 7 + 3 = 10, then you also know 70 + 30 = 100 and 700 + 300 = 1000.

Complete this table.

1	9 + 1 = 10	90 + 10 = ___	900 + 100 = ___
2	___ + 8 = 10	___ + ___ = ___	___ + ___ = ___
3	7 + ___ = ___	___ + ___ = ___	___ + ___ = ___
4	___ + 4 = ___	___ + ___ = ___	___ + ___ = ___
5	5 + ___ = ___	___ + ___ = ___	___ + ___ = ___

Knowing doubles is useful too.
If you know 7 + 7 = 14, then you also know 70 + 70 = 140 and 700 + 700 = 1400.

Complete this table.

6	6 + 6 = ___	60 + 60 = ___	600 + 600 = ___
7	7 + 7 = ___	___ + ___ = ___	___ + ___ = ___
8	8 + 8 = ___	___ + ___ = ___	___ + ___ = ___
9	9 + 9 = ___	___ + ___ = ___	___ + ___ = ___
10	11 + 11 = ___	___ + ___ = ___	___ + ___ = ___

SCORE /10 0-4 5-8 9-10

TERM 3

Measurement & Space

AC9M3M04

How Many Minutes?

Work out how many minutes to the given times.

1. How many minutes to half past 3? ___

2. How many minutes to 7 o'clock? ___

3. How many minutes to 12 o'clock? ___

4. How many minutes to 6 o'clock? ___

5. How many minutes to quarter to 8? ___

SCORE /5 0-2 3-4 5

Statistics & Probability

AC9M3ST01

Comparing Results

Bindi and Luis surveyed students at their schools to find out which pets were most common. Here are their results:

Bindi's Results

Pet	Year 3	Year 4	Year 5
Dogs	12	15	9
Cats	6	8	12
Fish	5	4	3
Birds	1	0	2

Luis's Results

Pet	Year 3	Year 4	Year 5
Dogs	13	12	14
Cats	8	7	9
Fish	4	0	8
Birds	2	0	0
Horses	3	4	5

1. What would be a good title for their tables?

   ____

2. Which is the most common pet? ______________
3. Whose school has the most cats as pets? ______________
4. Whose school's Year 3 has the most pets altogether? ______________
5. How many categories did each survey have?

   Bindi's survey: ____ Luis's survey: ____
6. How many more cats did the students at Bindi's school have than the students at Luis's school?

   ____
7. How many fewer dogs did the students at Bindi's school have than the students at Luis's school?

   ____

SCORE /7 0-3 4-5 6-7

NAPLAN*-style Questions

1. Write the number in the box that makes this number sentence true.

   70 + 30 = 60 + ☐
2. This clock shows the time at which Max wakes up.
   At what time does Max wake up?

   ☐

3. Kiku started with 14 marbles.
   At the end of the marbles tournament, she had double that number of marbles.
   How many marbles does Kiku have now?

   Kiku now has ☐ marbles.

SCORE /3 0-1 2 3

TARGETING MATHS HOMEWORK: YEAR 3 © PASCAL PRESS ISBN: 9781925726565

Problem of the Week

UNIT 21

AC9M3A02

Making Totals

"Well, class!" said Mr Green. "Here is my problem for you.

- "Choose three of the numbers 1, 2, 3, 4, 5 and 6 to make a total.
- You may use addition or a mixture of addition and subtraction.
- You must not use a number twice.
- You must use 3 numbers each time.

"Show me how to make the totals listed below with three numbers. Where possible, show more than one way of making that total. For example, 6 + 4 – 2 makes the total 8."

Here are the totals to make:

To make	Use	Or Use
1		
2		
3		
4		
5		
6		
7		
8		

To make	Use	Or Use
9		
10		
11		
12		
13		
14		
15		

"Was that too easy?" Mr Green asked. "If you would like to take my problem further, try including 7 and 8 in the list of numbers that can be used."

List the new totals that are possible in this table.

To make	Use	Or Use

To make	Use	Or Use

"You could go one step further and include the numbers 9 and 10 as well. That would be a real challenge!"

List the new totals that are possible in this table.

To make	Use	Or Use

To make	Use	Or Use

TERM 3

AC9M3A01

Unknowns in Subtraction

Complete these subtraction questions where the answer is unknown.

① 26 – 4 = ☐ ② 33 – 8 = ☐ ③ 42 – 6 = ☐

In the subtraction ☐ – 6 = 32, the first number or **start** number is **unknown**.
To solve this, we ask 32 + 6 = **what?**
The answer is 38 and that tells us that the **start** number in ☐ – 6 = 32 is **38**.

Use addition to find the start unknown in these examples.

④ ☐ – 6 = 30 ⑥ ☐ – 6 = 39 ⑧ ☐ – 8 = 52

⑤ ☐ – 7 = 28 ⑦ ☐ – 9 = 53 ⑨ ☐ – 6 = 25

In the subtraction 46 – ☐ = 39, the second number or **change** is **unknown**.
We do not know what was subtracted, so we ask 39 + what = 46.
The answer is 7 and that tells us that the **change** number in 46 – ☐ = 39 is **7**.

Use addition to find the change unknown in these examples.

⑩ 27 – ☐ = 21 ⑫ 27 – ☐ = 19 ⑭ 28 – ☐ = 18

⑪ 64 – ☐ = 58 ⑬ 72 – ☐ = 60 ⑮ 101 – ☐ = 99

TERM 3

SCORE /15 0-7 8-13 14-15

Quick Check

**Write the type of subtraction: answer, start or change unknown.
Then complete the subtraction.**

① ____________ unknown
37 – 3 = ☐

② ____________ unknown
28 – ☐ = 19

③ ____________ unknown
☐ – 6 = 54

④ ____________ unknown
103 – 4 = ☐

⑤ ____________ unknown
☐ – 7 = 63

⑥ ____________ unknown
☐ – 10 = 38

⑦ ____________ unknown
94 – 5 = ☐

⑧ ____________ unknown
73 – ☐ = 64

⑨ ____________ unknown
35 – ☐ = 21

⑩ ____________ unknown
1000 – ☐ = 996

SCORE /10 0-4 5-8 9-10

TARGETING MATHS HOMEWORK: YEAR 3 © PASCAL PRESS ISBN: 9781925726565

Pattens in Multiplication Tables

There is a pattern in the 5 times multiplication table. Continue this 5s counting sequence.

①

5	10	15	20							

② Is 167 in the 5s counting sequence? ◯ Yes ◯ No

③ How do you know? ____________________

④ Is 1365 in the 5s counting sequence? ◯ Yes ◯ No

⑤ How do you know? ____________________

There is a special feature in the 4 times multiplication table. Continue this 4s counting sequence.

⑥

4	8	12	16							

⑦ Is 73 in the 4s counting sequence? ◯ Yes ◯ No

⑧ How do you know? ____________________

⑨ Is 124 in the 4s counting sequence? ◯ Yes ◯ No

⑩ How do you know? ____________________

SCORE /10 0-4 5-8 9-10

Statistics & Probability

AC9M3ST02

Shapes in a Box

I have been sorting the shapes in the shape box.
This is the list I made: 16 hexagons, 8 diamonds, 26 squares, 24 circles, 28 triangles.
I have filled in the hexagon row on my chart. Please finish the chart for me.

① What key did Ada use?

One shape = ____

② Write a title for the graph.

____

____

③ Complete the graph.

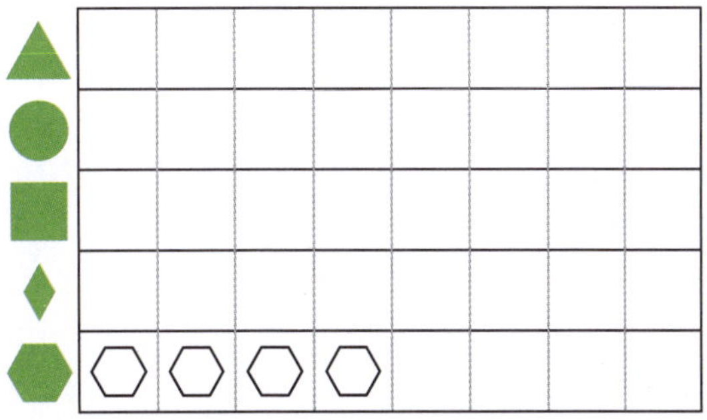

SCORE /6 0-2 3-4 5-6

AC9M3M01

Estimating Mass

Find something in your kitchen cupboard that has a mass of 500 g.

- Hold the object that you found and think about what it feels like to hold something that is 500 g.
- Now choose four other things that you can compare with your 500 g object. List them in the table below.
- Tick the box that matches your estimate.
- If you have kitchen scales, check your estimates.

Here are a few ideas to look at in your kitchen cupboard: packet of sugar, jar of pasta sauce, packet of coffee beans or a box of cereal.

	Object	Less than 500 g	About 500 g	More than 500 g	Actual Measured Mass
1					
2					
3					
4					

Now repeat the activity with an object that weighs 1 kg.

	Object	Less than 1 kg	About 1 kg	More than 1 kg	Actual Measured Mass
5					
6					
7					
8					

SCORE /8 0-3 4-6 7-8

NAPLAN*-style Questions

1. At the start of the school week Luis had $25 for school lunches.
He spent $4 every day on his lunch.
How much money did he have left at the end of the week?

   ◯ $4 ◯ $20 ◯ $5 ◯ $21

2. Max has more than twice as many stars as Kiku, but fewer than Hugo.
How many stars does Max have?

   ◯ 1 ◯ 2 ◯ 3 ◯ 3

Max	
Kiku	★
Hugo	★★★★
Ada	★★★★★

3. Which of these has a mass of about 500 g?

   ◯ 3 apples ◯ a cat
   ◯ an egg ◯ 1 litre of milk

SCORE /3 0-1 2 3

TARGETING MATHS HOMEWORK: YEAR 3 © PASCAL PRESS ISBN: 9781925726565

Going Camping

We are going on an overnight camping expedition and we have chosen the food to take with us.

We have to carry the food in four backpacks. To make sure that we each carry our share, all four backpacks must carry the same mass.

What should we put in each pack?

Milk 2 L

Milk 2 L

Orange Juice 1 L

Orange Juice 1 L

Apples 125 g each

Burgers 500 g

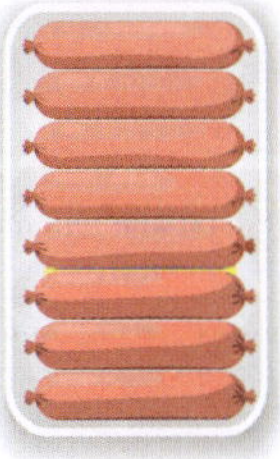
Sausages 625 g

Cereal 375 g each

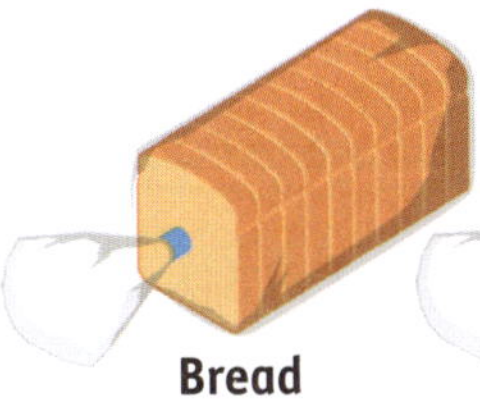
Bread 1 kg

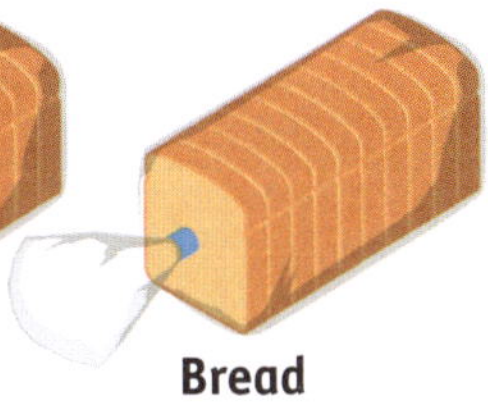
Bread 1 kg

Tomato Sauce 500 g

Tomatoes 800 g

Honey 200 g

List the items to go into each backpack.

Bindi's backpack

Max's backpack

Luis's backpack

Kiku's backpack

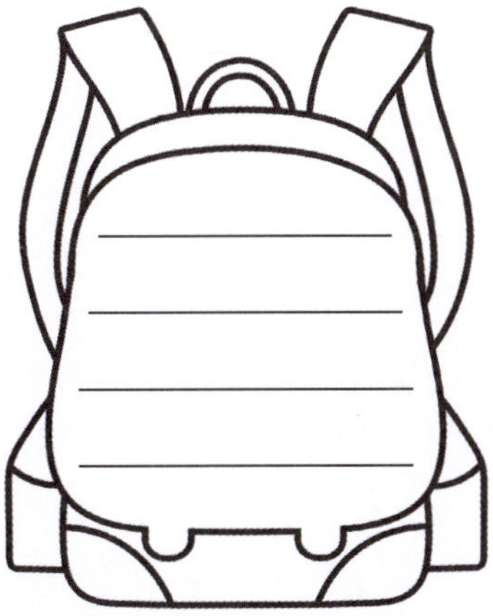

Number & Algebra

AC9M3N04

Multiplication

3 × 60 is like 3 × 6 × 10.

We know that 3 × 6 = 18 and that 18 × 10 = 180, so 3 × 60 = 180.

Complete these multiplications.

(1) 20 × 4 = 2 × 4 × 10 = ☐

(2) 30 × 5 = 3 × ☐ × ☐ = ☐

(3) 6 × 40 = ☐ × ☐ × ☐ = ☐

(4) 70 × 3 = ☐ × ☐ × ☐ = ☐

For 32 × 6, split the 32 into two parts, 30 + 2.

Multiply both parts by 3 and combine the results.

Complete these multiplications.

(5) 32 × 6

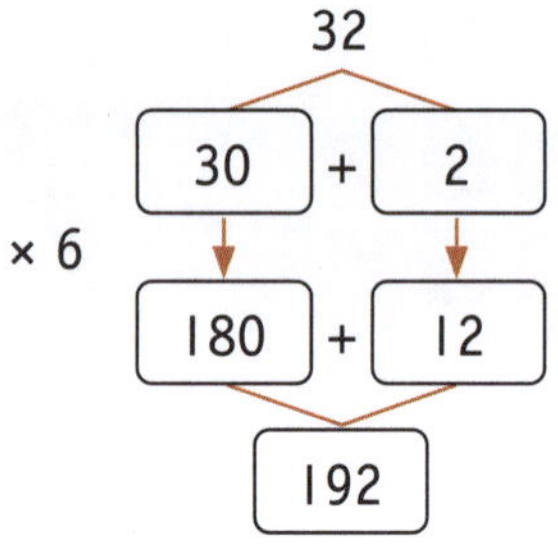

(7) 41 × 3

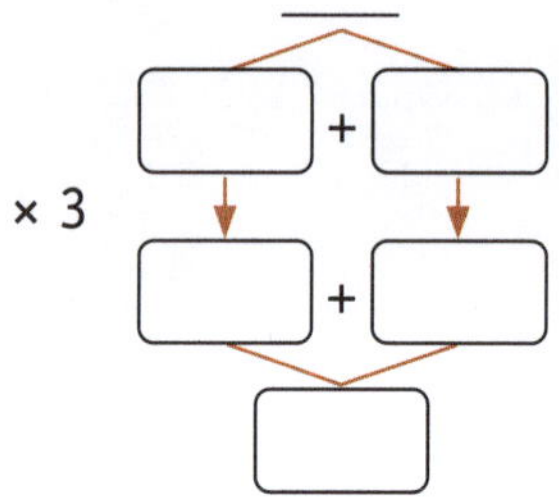

(9) 53 × 5

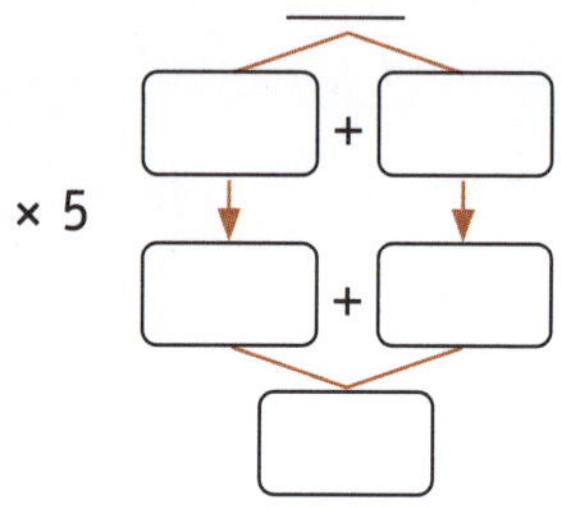

(6) 27 × 3

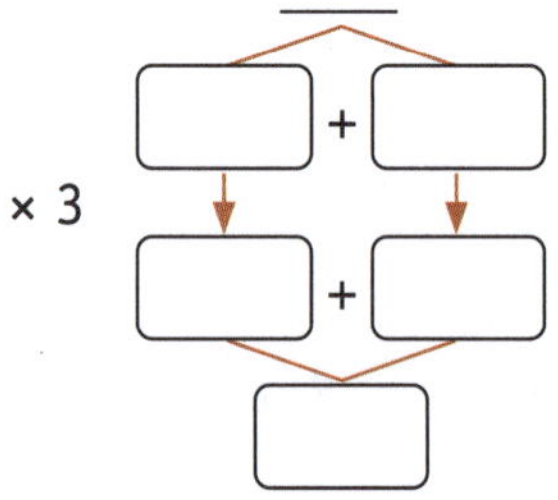

(8) 38 × 4

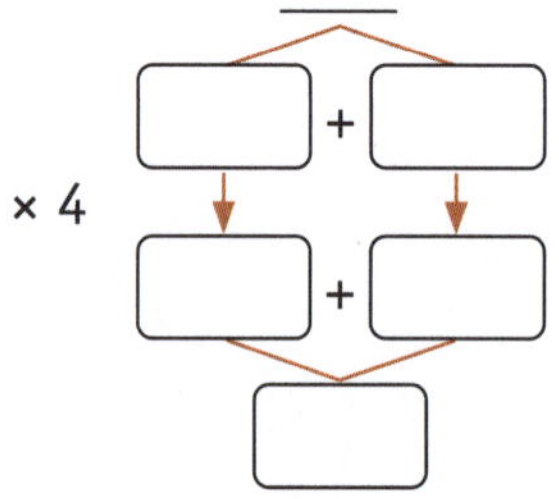

(10) 42 × 6

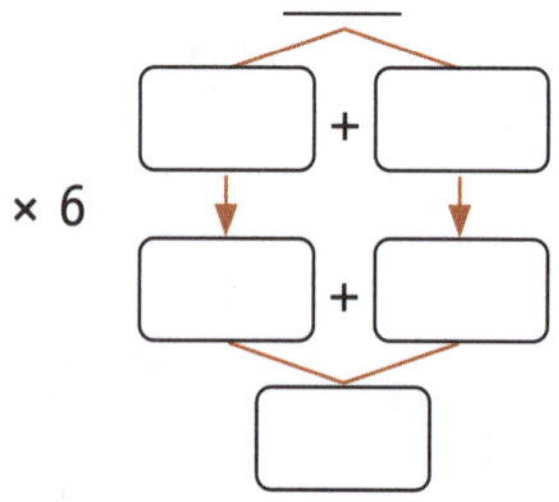

SCORE /10 0-4 5-8 9-10

Quick Check

Complete these multiplications.

(1) 30 × 4 = ______

(2) 50 × 5 = ______

(3) 70 × 3 = ______

(4) 80 × 4 = ______

(5) 110 × 4 = ______

(6) 31 × 3 = ______

(7) 42 × 4 = ______

(8) 51 × 5 = ______

(9) 47 × 6 = ______

(10) 92 × 10 = ______

SCORE /10 0-4 5-8 9-10

TARGETING MATHS HOMEWORK: YEAR 3 © PASCAL PRESS ISBN: 9781925726565

Multiplication Word Problems

Solve these multiplication word problems.

1. Bread rolls come in packets of 6.
   How many rolls are there in 4 packets?
   4 × 6 = ☐

2. Marker pens cost $3 each.
   How much would 8 marker pens cost?
   ☐ × ☐ = ☐

3. Hugo scored 6 points in each quarter of the basketball game.
   How many points did Hugo score altogether?
   ☐ × ☐ = ☐ points

4. Ada has 5 times as many stickers as Bindi who has 3 stickers.
   How many stickers does Ada have?
   ☐ × ☐ = ☐ stickers

5. Kiku read 24 pages of her book every school night this week.
   How many pages did Kiku read altogether?
   ☐ × ☐ = ☐ pages

6. Max runs for 10 minutes every day of the week.
   How much time does he spend running in one week?
   ☐ × ☐ = ☐ min

7. There are three times as many tadpoles in the pond as there are frogs.
   We counted 22 frogs altogether.
   How many tadpoles are there in the pond?
   ☐ × ☐ = ☐ tadpoles

SCORE /7 0-3 4-5 6-7

Making a Column Graph

Luis and Max counted the creatures they found in the school garden.
They made tallies to record their counts. Now they need to make a column graph of the data.

Luis and Max's Data	
Ants	𝍸 𝍸 𝍸 |
Bees	𝍸 |||
Spiders	𝍸 𝍸 𝍸 𝍸 ||||
Ladybugs	𝍸 𝍸 ||

Title: ____________________

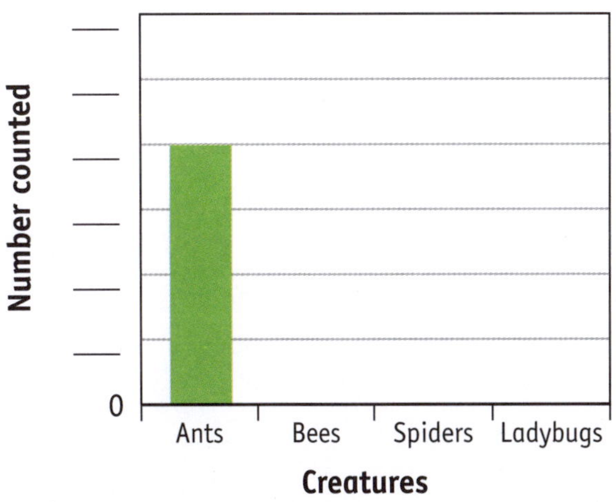

1. Write a title for the graph.
2. Write the missing numbers on the scale.
3. Complete the graph.

SCORE /3 0-1 2 3

UNIT 23

Measurement & Space

AC9M3SP02

A Lost Dog

Kiku's dog has escaped. The map shows where Kiku is and where the naughty dog is.
Write instructions for the shortest route that Kiku can take to collect her dog.
Use words such as north, south, east, west, left, right, turn and straight ahead.

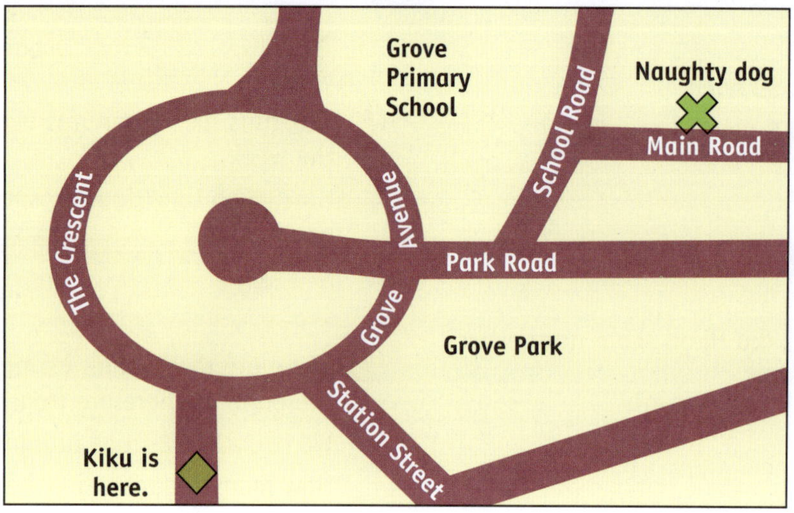

Directions

① ________________________________

② ________________________________

③ ________________________________

④ ________________________________

⑤ ________________________________

SCORE /5 0-2 3-4 5

NAPLAN*-style Questions

① There are 16 oranges in each bag.
How many oranges are there altogether?

- ⬭ 2
- ⬭ 16
- ⬭ 32
- ⬭ 18

② Look at the table.
How many more students prefer carrots than prefer beans?

- ⬭ 7
- ⬭ 1
- ⬭ 2
- ⬭ 3

Vegetable	Number of Votes	
	Girls	Boys
Carrots	6	8
Broccoli	8	4
Peas	16	12
Beans	3	9

③ Hugo's journey from school to home is 3 km.
How many kilometres does Hugo travel to and from school each week?

- ⬭ 6 km
- ⬭ 12 km
- ⬭ 11 km
- ⬭ 30 km

SCORE /3 0-1 2 3

TARGETING MATHS HOMEWORK: YEAR 3 © PASCAL PRESS ISBN: 9781925726565

Going Bush

Twenty-eight students and four adults are going on a bush excursion next week.

They are planning to have salad sandwiches, and then fresh fruit and iced water.

But Mr Green is not happy with their plan.

"Students! I know that is your first plan, but it needs fixing.
Look closely at your first plan and then complete the correct plan."

The First Plan	A Correct Plan
6 packets of 4 bread rolls	____ packets of 4 bread rolls
3 packets of 16 cheese slices (2 slices for each bread roll)	____ packets of 16 cheese slices (2 slices for each bread roll)
9 packets of 8 tomatoes (1 tomato for each bread roll)	____ packets of 8 tomatoes (1 tomato for each bread roll)
64 bottles of iced water (3 bottles per person)	____ bottles of iced water (3 bottles per person)
14 apples (half each)	____ apples (half each)
120 strawberries (6 strawberries each)	____ strawberries (6 strawberries each)
56 mandarins (2 each)	____ mandarins (2 each)

Now make up a sensible plan for a healthy lunch for your class.

Fractions and Division

Finding half is the same as dividing by 2.

1. Complete this **finding half** table.

Number	10	20	30	40	50	60	70	80	90	100
half ÷ 2	5	10	15							

Split the numbers and divide by 2.

2. 22 ÷ 2

22
20 + 2
÷ 2
10 + 1
11

3. 28 ÷ 2

___
☐ + ☐
÷ 2
☐ + ☐
☐

4. 36 ÷ 2

___
☐ + ☐
÷ 2
☐ + ☐
☐

Finding a third is the same as dividing by 3.

5. Complete this **finding a third** table.

Number	15	30	45	60	75	90	105	120	135	150
a third ÷ 3	5	10	15							

Split the numbers and divide by 3.

6. 36 ÷ 3

36
30 + 6
÷ 3
10 + 2
12

7. 63 ÷ 3

___
☐ + ☐
÷ 3
☐ + ☐
☐

8. 96 ÷ 3

___
☐ + ☐
÷ 3
☐ + ☐
☐

SCORE /8 0-3 4-6 7-8

Quick Check

Find half.

1. 16 ☐
2. 24 ☐
3. 46 ☐
4. 32 ☐
5. 54 ☐

Find one-third.

6. 30 ☐
7. 66 ☐
8. 39 ☐
9. 48 ☐
10. 72 ☐

SCORE /10 0-4 5-8 9-10

TERM 3

TARGETING MATHS HOMEWORK: YEAR 3 © PASCAL PRESS ISBN: 9781925726565

Division Word Problems

Solve these division word problems.

① There are 36 cherries.
Ada, Bindi and Luis each need one-third of the cherries for their fruit salad.
How many cherries is that?

________ = ____ cherries

② 27 oranges need to be put equally into 3 nets.
How many oranges should be put into each net?

________ = ____ oranges

③ Max has read half of his 96-page book.
How many pages does Max have to read to finish his book?

________ = ____ pages

④ Hugo has spent half of his $74 savings.
How much money has Hugo spent?

________ = $____

⑤ Kiku has one-third of her 96-page book left to read.
How many pages does Kiku have to read to finish her book?

________ = ____ pages

⑥ Ada has eaten one-third of the 24 strawberries.
How many strawberries has she eaten?

________ = ____ strawberries

⑦ Luis needs to share $12 equally between himself and his three sisters.
How much money should each person get?

________ = $____

SCORE /7 0-3 4-5 6-7

Statistics & Probability

Probability

Ada made this spinner.

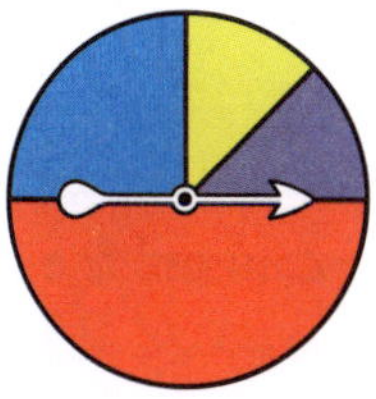

① Which colour would you expect the spinner to land on most often?

____

② Which two colours have the same chance of being spun?

________ and ________

Ada had 25 turns with her spinner and made this tally of the results.

Red	Blue	Yellow	Purple
𝍸 III	𝍸 IIII	III	𝍸

③ Which colour did the spinner land on most often? ________

④ Were two colours landed on the same number of times? ◯ Yes ◯ No

⑤ Were you surprised that red was not landed on most often? ◯ Yes ◯ No

⑥ Why? ________________________

SCORE /6 0-2 3-4 5-6

AC9M3M01, AC9M3M02

TERM 3

Half Full

Each of these jugs is half full. How much juice is in each jug?

① ________ ② ________ ③ ________ ④ ________

Each of these boxes of rice is half empty. How much rice has been taken from each box?

⑤ ________ ⑥ ________ ⑦ ________ ⑧ ________

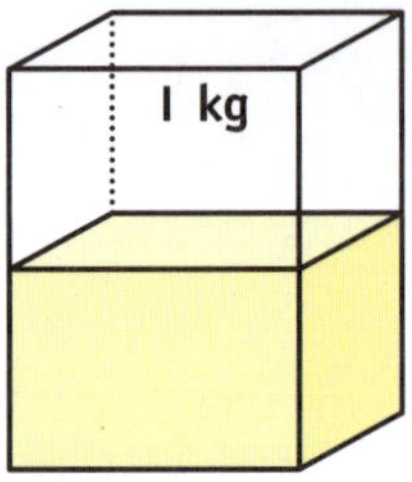

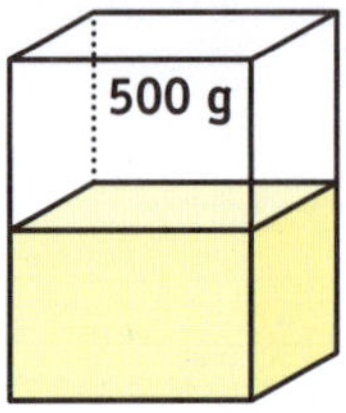

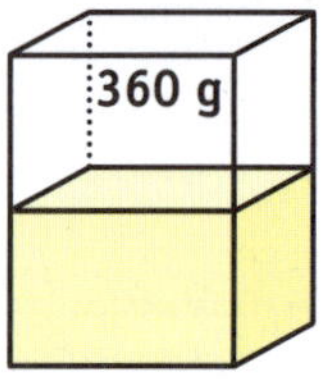

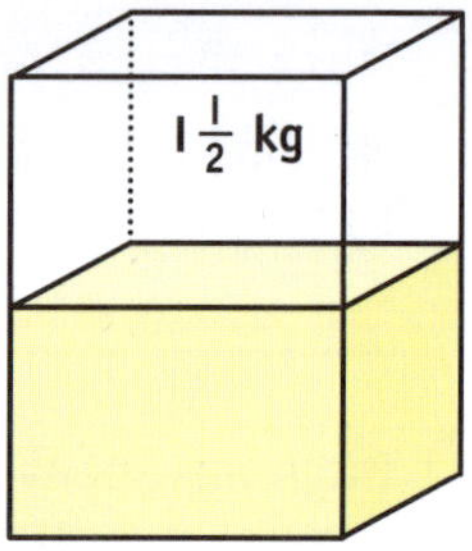

SCORE /8 0-3 4-6 7-8

NAPLAN*-style Questions

① There are 26 apples altogether, shared equally.
How many apples are in each bag?

☐

② Bindi counted the birds visiting the birdbath.
There were 24 altogether and one-third of them were sparrows.
How many sparrows visited the birdbath?

☐

③ Each of these boxes of rice is half full. Which box has the least rice in it?

◯ ◯ ◯ ◯

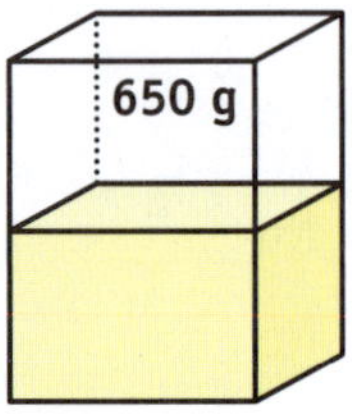

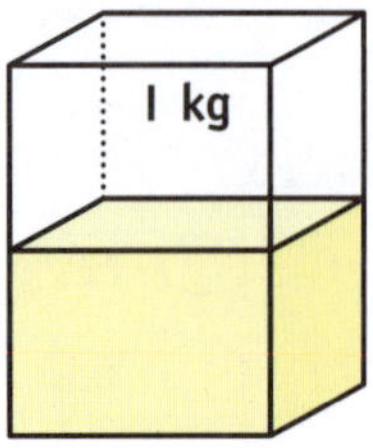

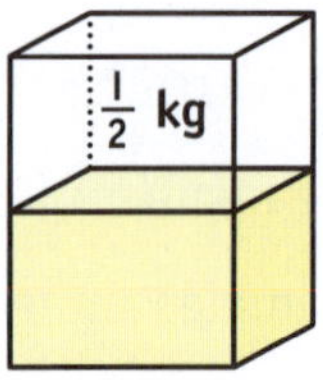

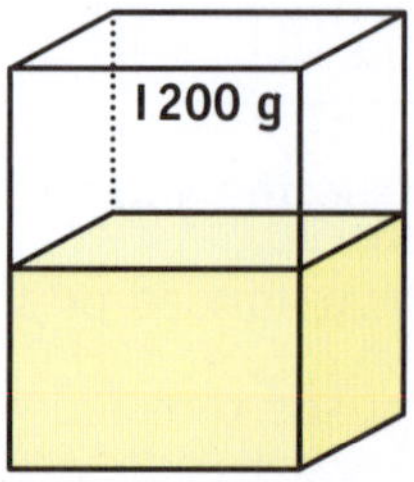

SCORE /3 0-1 2 3

TARGETING MATHS HOMEWORK: YEAR 3 © PASCAL PRESS ISBN: 9781925726565

Problem of the Week

UNIT 24

AC9M3N04

Guess the Number Puzzles (With Division)

Solve these puzzles.

1. Luis's number can be divided by 2 exactly.
   It can also be divided by 3 exactly.
   His number is more than 10 but less than 20 and it is closer to 20 than it is to 10.

   Luis's number is ☐.

2. Max's number can be divided by 2 exactly.
   It can also be divided by 5 exactly.
   It is the fifth number in the counting sequence of numbers that can be divided by 2 and 5 exactly.

   Max's number is ☐.

The next puzzles are more difficult. You may need to use Guess, Check and Improve strategy or the Work Backwards strategy using multiplication.

3. If you divide Kiku's number by 3 and then divide the answer by 2, the answer is 4.

   Kiku's number is ☐.

4. If you halve Hugo's number and then divide the answer by 5, the answer is 2.

   Hugo's number is ☐.

5. You can halve Ada's number and the answer is a whole number.
   You can also find one-third of Ada's number and the answer is a whole number.
   Her number is less than 12.

   Ada's number is ☐.

6. Hi, I'm Bindi and I haven't made up a number puzzle with division. I need your help!
   Write a number puzzle with division for Bindi.

   ____

   ____

7. Bindi's number is ☐.

TERM 3

TERM 3 REVIEW

Review: Unit 17 and Unit 18

AC9M3N02, AC9M3N06, AC9M3N07, AC9M3A01, AC9M3M02, AC9M3M06, AC9M3ST01

Continue these pattern sequences.

(1) 3, 7, 11, ____, ____, ____, ____, ____, ____

(2) 12, 15, 18, ____, ____, ____, ____, ____, ____

(3) 2, 4, 3, 5, ____, ____, ____, ____, ____

(4) 12, 23, 34, ____, ____, ____, ____, ____, ____

(5) 91, 96, 95, ____, 99, 104, ____, ____, ____

Complete these answer unknown problems.

(6) 16 + 11 = ☐ (7) 25 + 13 = ☐ (8) 64 + 37 = ☐

Complete these start unknown problems.

(9) ☐ + 15 = 46 (10) ☐ + 25 = 67 (11) ☐ + 37 = 108

(12) This graph shows the coins that Luis had collected. How much money has Luis collected altogether?

____

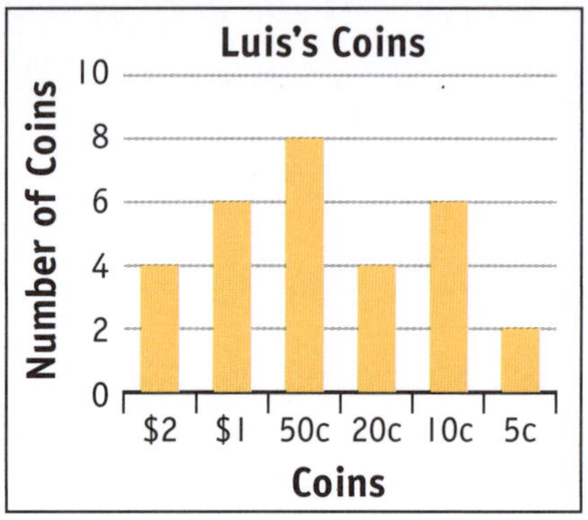

(13) This line is 3 cm long.

How long is this line?

____ cm

TERM 3

Each of these pieces is a fraction of a whole circle. Write the fraction for each piece.

(14) 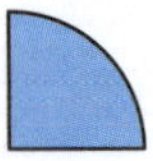 ☐

(15) 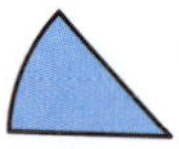 ☐

(16) 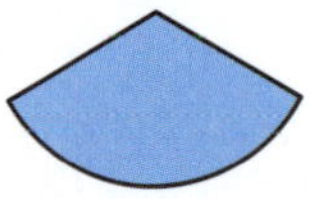 ☐

(17) 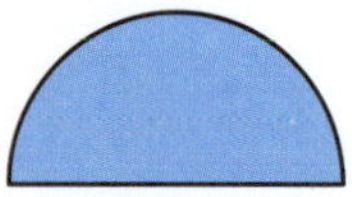 ☐

Complete these change unknown problems.

(18) 20 + ☐ = 63 (19) 56 + ☐ = 68 (20) 95 + ☐ = 102

Show how these coins can be sorted into groups such that there is the same amount of money in each group and the same number of coins in each group.

(21) Group 1: ____________ (22) Group 2: ____________ (23) Group 3: ____________

SCORE /23 0-12 13-21 22-23

TARGETING MATHS HOMEWORK: YEAR 3 © PASCAL PRESS ISBN: 9781925726565

Review: Unit 19 and Unit 20

AC9M3N01, AC9M3A03, AC9M3M03

Complete these multiplication facts.

(1) 4 × 3 = ☐

(2) 6 × 3 = ☐

(3) 5 × 7 = ☐

(4) 4 × 8 = ☐

(5) 3 × 6 = ☐

(6) 7 × 5 = ☐

(7) 12 × 6 = ☐

(8) 11 × 5 = ☐

Write two multiplication facts to match this array.

(9) ☐ × ☐ = ☐

(10) ☐ × ☐ = ☐

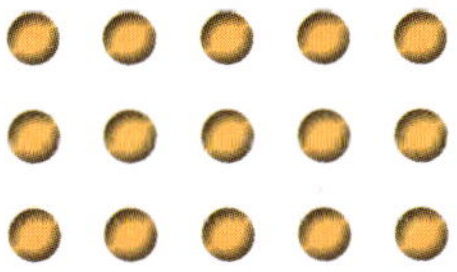

Use the calendar to answer the following questions.

(11) What day of the week is 21st June?

____

(12) What is the date of the first Saturday in June?

____

(13) What is the date of the last Sunday in June?

____

(14) What is the next month after June?

____

June						
Sun.	Mon.	Tue.	Wed.	Thu.	Fri.	Sat.
28	29	30	31	1	2	3
4	5	6	7	8	9	10
11	12	13	14	15	16	17
18	19	20	21	22	23	24
25	26	27	28	29	30	1
2	3	4	5	6	7	8

How much are these play money notes worth?

(15) 4 hundreds, 5 tens and 11 ones

$__________

(16) 12 hundreds and 21 ones

$__________

(17) 2 thousands, 6 hundreds and 15 ones

$__________

(18) 3 thousands, 9 hundreds and 11 tens

$__________

(19) 20 thousands, 6 hundreds and 6 ones

$__________

Use friendly numbers to complete these additions.

(20) $28^{+2} + 53^{-2}$

☐ + ☐ = ☐

(21) $36^{+\_\_} + 45^{-\_\_}$

☐ + ☐ = ☐

(22) $78^{+\_\_} + 16^{-\_\_}$

☐ + ☐ = ☐

(23) $69^{+\_\_} + 72^{-\_\_}$

☐ + ☐ = ☐

(24) $96^{+\_\_} + 25^{-\_\_}$

☐ + ☐ = ☐

(25) $77^{+\_\_} + 33^{-\_\_}$

☐ + ☐ = ☐

SCORE /25 0-12 13-23 24-25

TERM 3

TERM 3 REVIEW

Review: Unit 21 and Unit 22

AC9M3N05, AC9M3M01, AC9M3M04, AC9M3A01, AC9M3ST01

Show how to make a reasonable estimate of the answer to these additions.

1. For 39 + 33, use 40 + 33 = ☐
2. For 27 + 34, use ☐ + ☐ = ☐
3. For 23 + 59, use ☐ + ☐ = ☐
4. For 68 + 34, use ☐ + ☐ = ☐
5. For 99 + 23, use ☐ + ☐ = ☐
6. For 86 + 76, use ☐ + ☐ = ☐

Work out how many minutes to the given times.

7. 

How many minutes to 4 o'clock? _____

8. 

How many minutes to 7 o'clock? _____

9. 

How many minutes to 7:05? _____

The tables show the flower shop sales for Monday and Tuesday.

Monday	
Roses	15
Tulips	28
Daffodils	36
Gladioli	12

Tuesday	
Roses	24
Dahlias	16
Daffodils	24
Asters	26

10. On which day were most roses sold? _______________
11. On which day were fewest flowers sold? _______________
12. How many more daffodils were sold on Monday than on Tuesday? ____
13. Which flower sold the most over the two days? _______________
14. Which flower sold the least over the two days? _______________

Complete these subtractions.

15. ☐ − 6 = 40
16. ☐ − 10 = 27
17. ☐ − 19 = 19
18. ☐ − 30 = 81
19. 26 − ☐ = 9
20. 48 − ☐ = 36

21. **Name 5 objects that have a mass of about 500 g.**

____

____

____

____

____

SCORE /21 0-10 11-19 20-21

TARGETING MATHS HOMEWORK: YEAR 3 © PASCAL PRESS ISBN: 9781925726565

TERM 3

Review: Unit 23 and Unit 24

AC9M3N02, AC9M3N04, AC9M3M01, AC9M3M02, AC9M3ST02

Complete these multiplications.

1. 20 × 3 = 2 × 3 × 10 = ☐
2. 30 × 3 = ☐ × ☐ × ☐ = ☐
3. 30 × 6 = ☐ × ☐ × ☐ = ☐
4. 40 × 4 = ☐ × ☐ × ☐ = ☐

Use number splitting to complete these multiplications.

5. 33 × 6

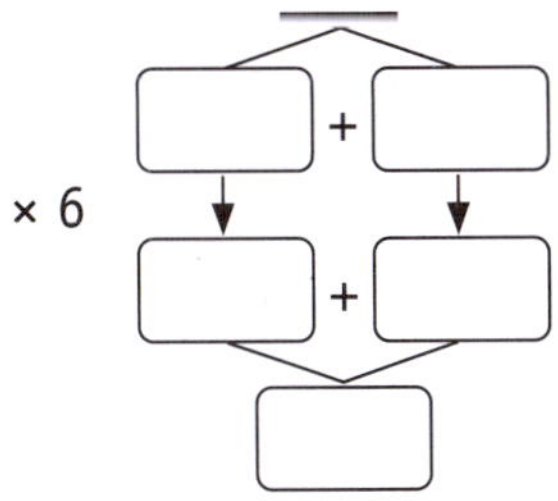

6. 25 × 3

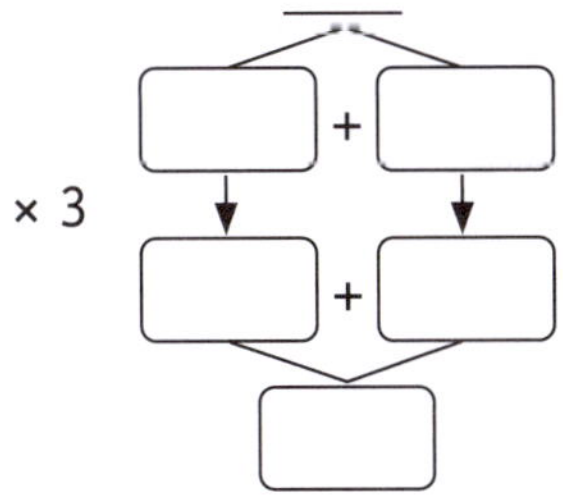

7. 15 × 7

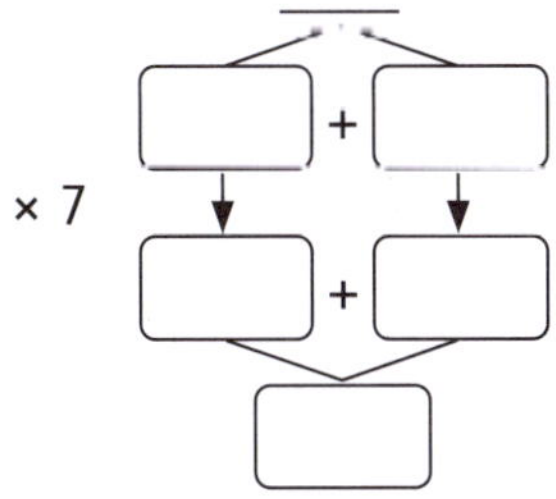

This is Max's creature graph.

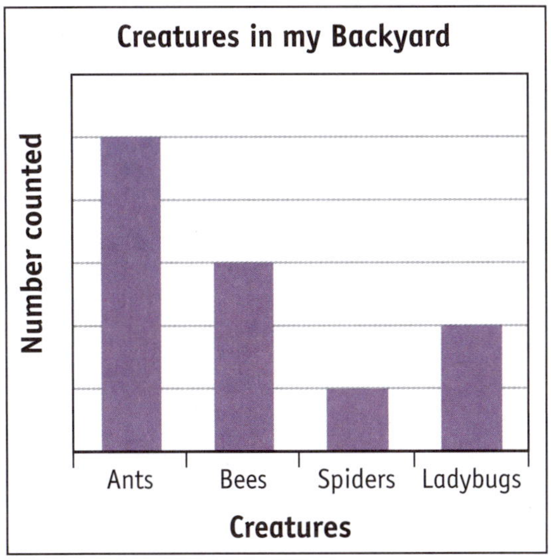

8. There were 5 spiders.
   What scale did Max use for each row on his graph?

   ____

9. How many ladybugs were there? ____
10. How many more bees than spiders were there? ____
11. How many creatures did he see altogether? ____

Solve these division word problems.

12. Max has 6 packets of stickers, each with 6 stickers.
    How many stickers does Max have altogether?

    ________________ stickers

13. Kiku shared her 24 superhero stickers equally between herself and three friends.
    How many stickers did each person have?

    ________________ stickers

Use number splitting to complete these divisions.

14. 86 ÷ 2

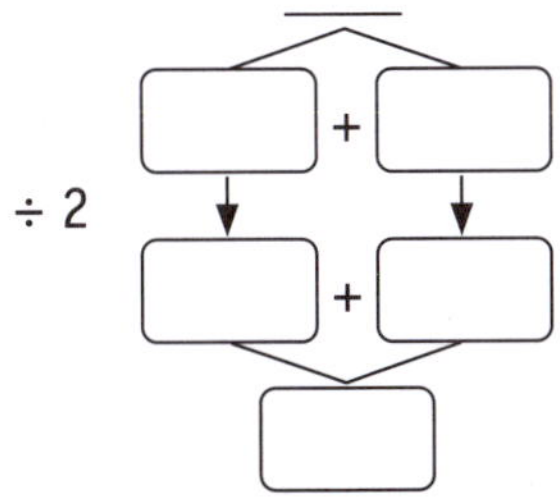

15. 63 ÷ 3

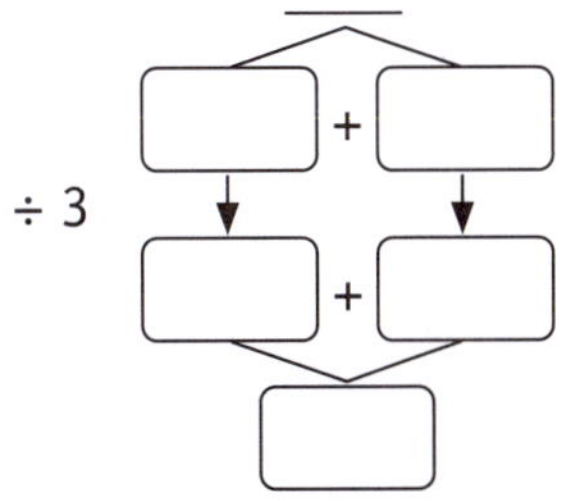

16. 48 ÷ 4

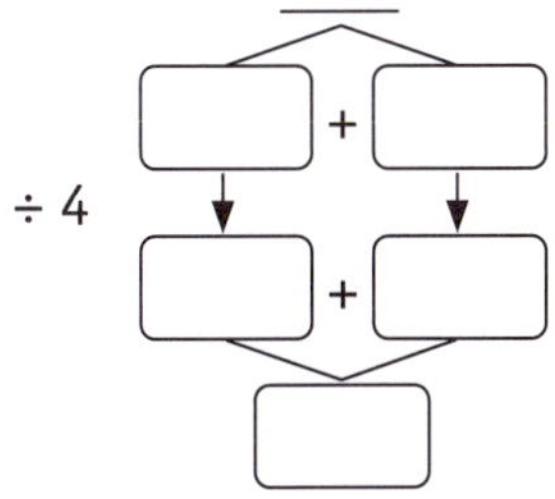

17. Which jug has the most liquid in?

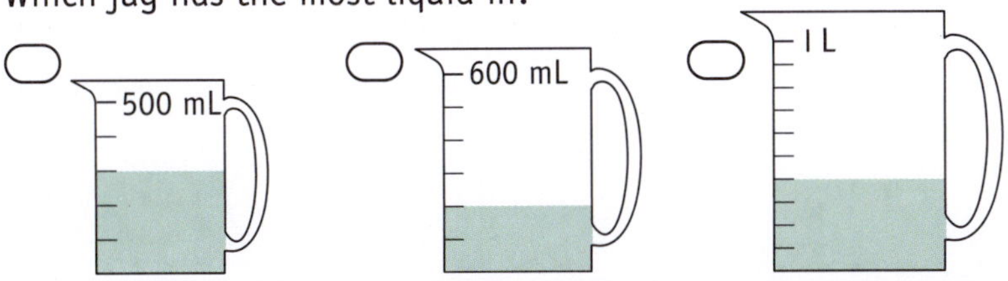

SCORE /17 0-8 9-15 16-17

Large Numbers

To read large numbers, look to see how many thousands there are.
Then read the last three digits to find the hundreds, tens and units.

12 437

[ ] thousand [ ]

The thousands are separated from the hundreds by a space.

12 437 is **twelve** thousand, **four** hundred and **thirty-seven**.

Complete the numbers.

1. **26 327**

   [ ] thousand [ ]

   In words: ______________ thousand, ________ hundred and ______________

2. **38 546**

   [ ] thousand [ ]

   In words: ______________ thousand, ________ hundred and ______________

3. **27 689**

   [ ] thousand [ ]

   In words: ______________ thousand, ________ hundred and ______________

How many thousands are there in these numbers?

4. 39 676 [ ] thousands
5. 43 279 [ ] thousands
6. 29 354 [ ] thousands
7. 60 000 [ ] thousands

SCORE /7 0-3 4-5 6-7

TERM 4

Quick Check

Write these numbers in numerals.

1. twenty-six thousand, four hundred and seventy-two [ ]
2. thirty-eight thousand, two hundred and eight [ ]
3. forty thousand, six hundred and fifty [ ]
4. fifty-five thousand and fifty [ ]
5. one hundred thousand and one [ ]

SCORE /5 0-2 3-4 5

TARGETING MATHS HOMEWORK: YEAR 3 © PASCAL PRESS ISBN: 9781925726565

Number & Algebra

AC9M3N07

Number Patterns

Follow the number pattern rules to continue these sequences. Then work out the 10th number in each sequence.

1. Start at 3, count on 3, subtract 1.

   3, 6, 5, ____, ____, ____, ____, ____ The 10th number is ____.

2. Start at 2, double, subtract 1.

   2, ____, ____, ____, ____, ____, ____, ____ The 10th number is ____.

3. Start at 51, subtract 6, add 2.

   51, ____, ____, ____, ____, ____, ____, ____ The 10th number is ____.

4. Start at 2, multiply by 3, subtract 3.

   2, ____, ____, ____, ____, ____, ____, ____ The 10th number is ____.

5. Start at 90, add 5, subtract 1.

   90, ____, ____, ____, ____, ____, ____, ____ The 10th number is ____.

6. Start at 110, subtract 5, add 3.

   110, ____, ____, ____, ____, ____, ____, ____ The 10th number is ____.

Write the rules for these sequences.

7. 26, 28, 24, 26, 22, 24, 20, ... Rule: ____________________
8. 3, 6, 4, 8, 6, 12, 10, 20, ... Rule: ____________________
9. 2, 3, 5, 8, 12, 17, 23, ... Rule: ____________________
10. 49, 47, 51, 49, 53, 51, 55, ... Rule: ____________________

SCORE /10 0-4 5-8 9-10

Statistics & Probability

AC9M3P01

What's the Weather?

Use the symbols and temperatures in this weather chart to predict the weather for the weekend.

Mon.	Tue.	Wed.	Thu.	Fri.

23 °C	20 °C	21 °C	20 °C	21 °C

1. What would you predict the weather to be on Saturday? ____________________
2. What would you predict the weather to be on Sunday? ____________________
3. What is the chance of rain at the weekend? ____________________
4. What is the chance of a 30° top temperature at the weekend? ____________________
5. What is the chance of snow at the weekend? ____________________

SCORE /5 0-2 3-4 5

TERM 4

AC9M3M04

Time Lines

Using a timeline helps when you are working out how long between a **start time** and a **finish time**.

For a start time of 3:45 and a finish time of 4:30, go from 3:45 to 4:00 and then on to 4:30.

15 min + 30 min = 45 min

Start time 3:45 — 4:00 — 4:30 Finish time

Use the timelines to work out how long between the start time and the finish time.

1. Start time 3:20 — 4:10 Finish time = ___ min
2. Start time 6:15 — 7:05 Finish time = ___ min
3. Start time 8:55 — 9:15 Finish time = ___ min
4. Start time 2:30 — 3:15 Finish time = ___ min
5. Start time 12:50 — 1:30 Finish time = ___ min

SCORE /5 0-2 3-4 5

NAPLAN*-style Questions

1. Bindi has shown a pattern rule on this number line.
What number came before 28 in Bindi's sequence?

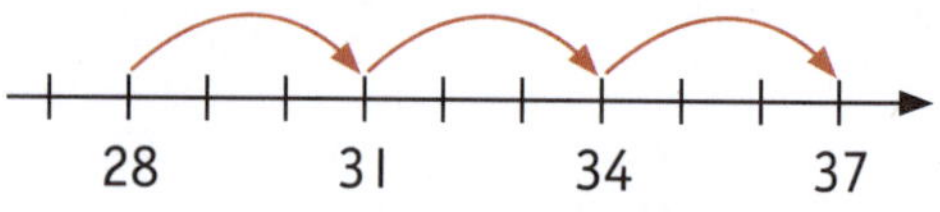

◯ 29 ◯ 27 ◯ 25 ◯ 40

2. Luis made the largest number possible using these number cards.
What number did Luis make?

6 0 7 9 5

____

3. Look at the clock.
What will the time be in a quarter of an hour?

◯ Quarter to eight ◯ 6:30
◯ 7:45 ◯ 4:00

SCORE /3 0-1 2 3

TARGETING MATHS HOMEWORK: YEAR 3 © PASCAL PRESS ISBN: 9781925726565

Problem of the Week

Screen Time

Max, Kiku and Hugo are cutting their evening screen time to 4 hours a week.

They each need a schedule for their screen times.

- Max is busy after school on Tuesdays and Wednesdays and has more spare time at weekends than on weeknights.
- Kiku wants less screen time at weekends and is out on Monday evenings.
- Hugo wants the same amount of screen time every evening of the week except Thursdays when he is out.

Use the tables below to create their viewing schedules.

There is more than one possible solution so choose the solutions you would prefer.

Max's Plan

Mon.	Tue.	Wed.	Thu.	Fri.	Sat.	Sun.	Total
							= 4 h

Kiku's Plan

Mon.	Tue.	Wed.	Thu.	Fri.	Sat.	Sun.	Total
							= 4 h

Hugo's Plan

Mon.	Tue.	Wed.	Thu.	Fri.	Sat.	Sun.	Total
							= 4 h

Do you have a plan for your own screen time?

Your Plan

Mon.	Tue.	Wed.	Thu.	Fri.	Sat.	Sun.	Total
							= 4 h

Fractions

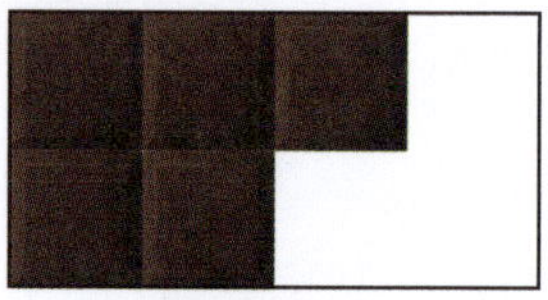

① What fraction of the chocolate bar has been eaten? ____

② What fraction of the chocolate bar is left over? ____

③ What fraction of the pizza has been eaten? ____

④ What fraction of the pizza is left over? ____

⑤ What fraction of the pizza has been eaten? ____

⑥ What fraction of the pizza is left over? ____

⑦ What fraction of the chocolate bar has been eaten? ____

⑧ What fraction of the chocolate bar is left? ____

⑨ What fraction of the eggs have been used? ____

⑩ What fraction of the eggs are left in the box? ____

⑪ Three apples were cut in halves. How many pieces of apple were there?

____ pieces of apple

⑫ Two oranges were cut into quarters. How many pieces of orange were there?

____ pieces of orange

⑬ Three cakes were cut into sixths. How many pieces of cake were there?

____ pieces of cake

SCORE /13 0-6 7-11 12-13

Quick Check

How much pizza is left?

① $\frac{1}{2}$ was eaten. ___

② $\frac{1}{4}$ was eaten. ___

③ $\frac{1}{8}$ was eaten. ___

④ $\frac{1}{3}$ was eaten. ___

⑤ $\frac{1}{5}$ was eaten. ___

How much pizza was eaten?

⑥ $\frac{3}{4}$ is left. ___

⑦ $\frac{3}{8}$ is left. ___

⑧ $\frac{1}{6}$ is left. ___

⑨ $\frac{2}{5}$ is left. ___

⑩ $\frac{7}{8}$ is left. ___

SCORE /10 0-4 5-8 9-10

TARGETING MATHS HOMEWORK: YEAR 3 © PASCAL PRESS ISBN: 9781925726565

Bar Model for Fractions

My friends, Ada and Bindi, have twenty stickers between them. Ada has four more stickers than Bindi.

I can use a bar model to work out how many stickers they each have.

First, I draw a bar for twenty stickers.

20

Next I show the four extra that Ada has.

4	16

Then I find half of sixteen to show how many stickers each of my friends has.

4	8	8

Ada's stickers (4 and 8) **Bindi's stickers** (8)

Altogether, the bar model looks like this:

20		
4	16	
4	8	8

Complete the bar model to answer these questions.

1. Luis has twelve more pencils than Max. Together they have twenty-four pencils. How many pencils do they each have?

   Luis has _____ pencils.

   Max has _____ pencils.

2. Kiku has eight fewer stickers than Hugo. Together they have twenty-two stickers. How many stickers do they each have?

   Kiku has _____ stickers.

   Hugo has _____ stickers.

3. Ada and Bindi have thirty pencils altogether. Bindi has six more pencils than Ada. How many pencils do they each have?

   Bindi has _____ pencils.

   Ada has _____ pencils.

4. Of the twenty-six crayons, Max has six more than Kiku. How many crayons do they each have?

   Max has _____ crayons.

   Kiku has _____ crayons.

5. Hugo has 30 more marbles than Ada. Together they have 64 marbles. How many marbles do they each have?

   Hugo has _____ marbles.

   Ada has _____ marbles.

SCORE /5 0-2 3-4 5

TARGETING MATHS HOMEWORK: YEAR 3 © PASCAL PRESS ISBN: 9781925726565

UNIT 26

Statistics & Probability

AC9M3ST02

Make a Chart

Bindi coloured this pie graph to show the results of spinning her spinner 24 times. Show Bindi's data on this bar graph.

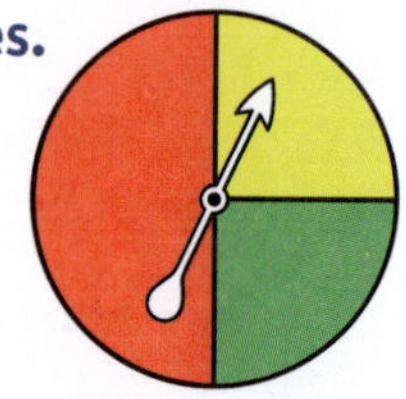

Red							
Yellow							
Green							

☐ = 2 spins

SCORE /3 0-1 2 3

Measurement & Space

AC9M3M02

Perimeter

1. Estimate the distance around (perimeter of) this page.

   Estimate: _____ cm

2. Check your estimate of the page's perimeter with a ruler.

   Perimeter = _____ cm

SCORE /2 0 1 2

NAPLAN*-style Questions

1. Luis, Max and Kiku share a packet of stickers.
   Luis and Max each get one-fifth of the stickers.
   What fraction of the stickers does Kiku get?

   ☐

2. Ada and Hugo picked 25 apples.
   Ada picked 5 more than Hugo.
   How many apples did Ada pick?

   ☐

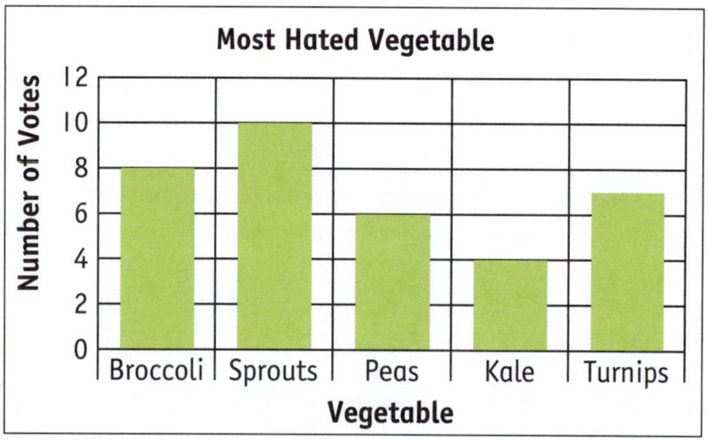

3. Bindi collected data from her classmates and made this graph.
   Which statement is true?

   - ◯ More students hate turnips than broccoli.
   - ◯ Fewer students hate peas than kale.
   - ◯ Sprouts are not the most hated vegetable.
   - ◯ The same number of students hate kale or peas as hate sprouts.

SCORE /3 0-1 2 3

TERM 4

TARGETING MATHS HOMEWORK: YEAR 3 © PASCAL PRESS ISBN: 9781925726565

The Giant Chocolate Bar

The giant chocolate bar was on sale for $36.

Luis, Max, Kiku and Hugo pooled their money to buy the giant chocolate bar.

- Luis paid $18.
- Max paid $12.
- Kiku and Hugo both paid the same amount.

1. How much did Kiku and Hugo both spend?

   Kiku and Hugo both spent $_____.

2. What fraction of the giant chocolate bar did Luis pay for?

   Luis paid for ____ of the bar.

3. What fraction of the giant chocolate bar did Max pay for?

   Max paid for ____ of the bar.

4. What fraction of the giant chocolate bar did Kiku and Hugo pay for?

   Kiku and Hugo each paid for ____ of the bar.

5. Use the diagram to show how many pieces of the giant chocolate bar each person paid for.

Write the totals here.

Luis had ____ pieces. Max had ____ pieces. Kiku had ____ pieces. Hugo had ____ pieces.

AC9M3N05

Estimation

Sometimes we do not need an exact, correct answer.

Instead, we need a **reasonable estimate**.

Use your estimation skills to say how many counters are hidden and why you chose that estimate.

(1)	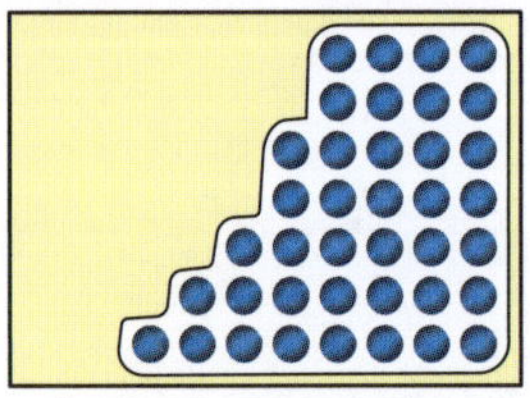	About ____ counters are hidden because ____________________
(2)	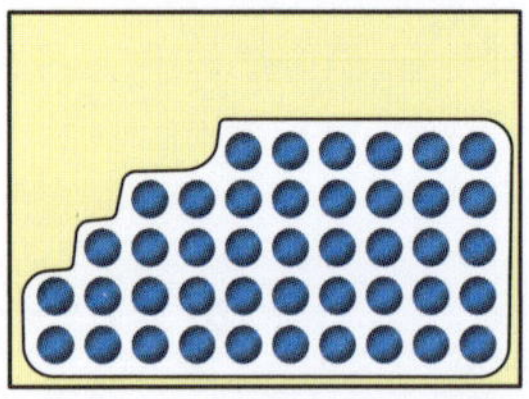	About ____ counters are hidden because ____________________
(3)	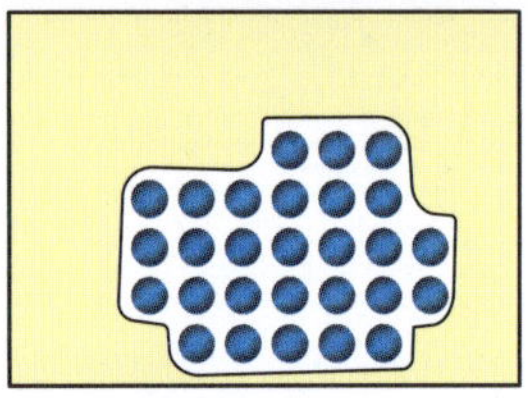	About ____ counters are hidden because ____________________
(4)	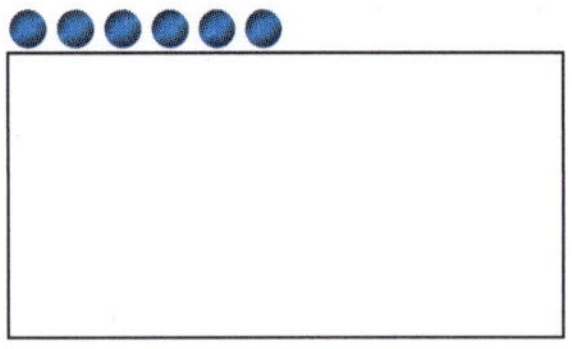	You need about ____ counters to go all around the edge (perimeter) of this rectangle because ____________________

SCORE /4 0-1 2-3 4

Quick Check

Estimate these amounts to the nearest $100.

(1) $694 ______ (2) $387 ______ (3) $405 ______ (4) $576 ______ (5) $908 ______

Estimate these amounts to the nearest $1000.

(6) $6985 ______ (7) $1876 ______ (8) $4808 ______ (9) $4067 ______ (10) $3154 ______

Estimate these amounts to the nearest $10 000.

(11) $9999 ______ (12) $39 888 ______ (13) $58 956 ______ (14) $40 875 ______ (15) $11 006 ______

SCORE /15 0-7 8-13 14-15

TERM 4

TARGETING MATHS HOMEWORK: YEAR 3 © PASCAL PRESS ISBN: 9781925726565

Number & Algebra

AC9M3N06

Number Sentences for Word Problems

Write the number sentences and work out the answers.

1. Bindi had $53 left after spending $36 at the shops.
   How much money did Bindi have to start with?

   Number sentence: ____________________________

2. Luis ran for 30 minutes every day this week.
   How many minutes did Luis run for altogether?

   Number sentence: ____________________________

3. 15 apples were packed equally into 3 nets.
   How many apples were in each net?

   Number sentence: ____________________________

4. 36 sausage rolls were put out for morning tea.
   9 sausage rolls were left over.
   How many sausage rolls were eaten?

   Number sentence: ____________________________

5. The dance teacher needs 6 lengths of elastic each 22 cm long for her costumes.
   How much elastic does she need altogether?

   Number sentence: ____________________________

SCORE /5 0-2 3-4 5

Statistics & Probability

AC9M3ST01

Dog-walking Income

Max walks his neighbours' dogs and records his income in a weekly graph.
The table shows his dog-walking income for last week.

Mon.	Tue.	Wed.	Thu.	Fri.	Sat.	Sun.
$15	$5	$10	$7·50	$5	–	$25

1. Complete Max's graph.

2. What is the scale on Max's graph?

   ☐ = $___

3. On which two days combined did Max make as much as he did on Sunday?

   ______________ and ______________

4. On which three days combined did Max make as much as he did on Sunday?

   ______________, ______________

   and ______________

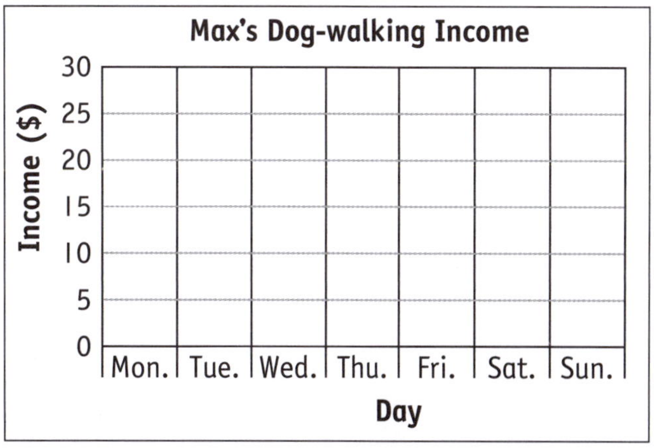

5. How much money did Max make for the whole week?

   Max made $________ for the whole week.

SCORE /5 0-2 3-4 5

TARGETING MATHS HOMEWORK: YEAR 3 © PASCAL PRESS ISBN: 9781925726565

AC9M3SP01

3D Objects and Nets

Match each object with the net that it is made from and find the net that has no matching object.

A B C D E

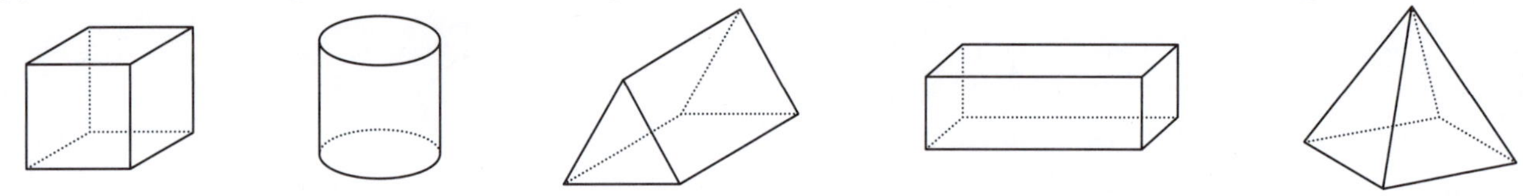

1 2 3 4 5 6

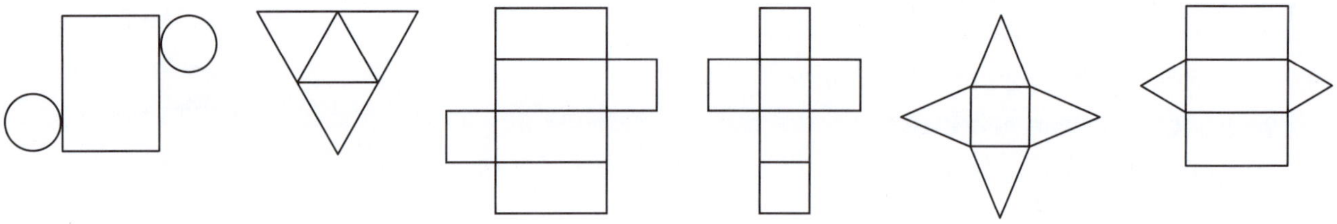

1. Object A is made with net ___.
2. Object B is made with net ___.
3. Object C is made with net ___.
4. Object D is made with net ___.
5. Object E is made with net ___.
6. Net ___ has no matching object.

SCORE /6 0-2 3-4 5-6

NAPLAN*-style Questions

1. What is the nearest thousand to 2368?

   ◯ 2000 ◯ 2300
   ◯ 2400 ◯ 23 000

2. Look at the graph.
   How many students travel 2 km or more but less than 3 km to school?

   ◯ 2
   ◯ 4
   ◯ 20
   ◯ 40

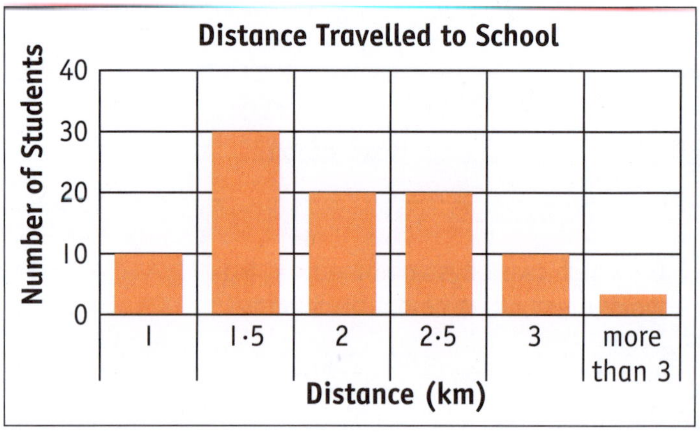

3. Kiku invited some of her friends to a barbecue.
   Seven friends came to the barbecue but three couldn't make it.
   How many friends did Kiku invite to her barbecue?

   ◯ 7 ◯ 10 ◯ 4 ◯ 13

SCORE /3 0-1 2 3

TERM 4

TARGETING MATHS HOMEWORK: YEAR 3 © PASCAL PRESS ISBN: 9781925726565

Problem of the Week

AC9M3ST01

The School Litter Problem

Mr Green's class were doing a litter survey.

Three teams covered three different areas of the school grounds.

Here are the Day 1 litter results:

Team 1
16 food wrappers
12 recyclable tubs
16 food scraps
2 lost property items
14 pieces of scrap paper

Team 2	
Food scraps	𝍷𝍷𝍷
Recyclables	𝍸 𝍷𝍷
Scrap paper	𝍸 𝍸 𝍷
Lost property	𝍷𝍷𝍷
Other	𝍷𝍷𝍷𝍷

Team 3	
Lost property	4
Food scraps	12
Scrap paper	6
Recyclables	8
Food wrappers	10

"This won't do!" said Mr Green.

"You need to put your data together so that you can answer these questions."

Mr Green wrote this list of questions on the board:

> How much litter was dropped altogether?
> How many:
> - food wrappers?
> - recyclables?
> - food scraps?
> - pieces of scrap paper?
> - items of lost property?
>
> What is the biggest litter offender?

Sort the teams' data and complete the table to show the totals in each of Mr Green's categories.

Rubbish items	Team 1	Team 2	Team 3	Total

Now answer Mr Green's questions.

How much litter was dropped altogether? ______

How many:

- food wrappers? ____
- recyclables? ____
- food scraps? ____
- pieces of scrap paper? ____
- items of lost property? ____

What is the biggest litter offender? ________________

Fraction Number Lines

If you count the equal spaces on a 0 to 1 number line, it tells you the fraction it has been divided into.

This number line is divided into thirds.

You can count on from 0 in thirds to label the positions on the line: zero, one-third, two-thirds and three-thirds which is 1.

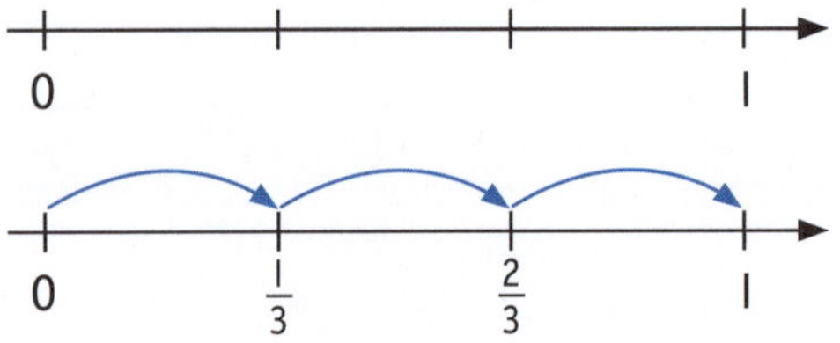

Count on and write the fraction you end on in number form. The first one has been done for you.

1. Start at 0 and count on one half.

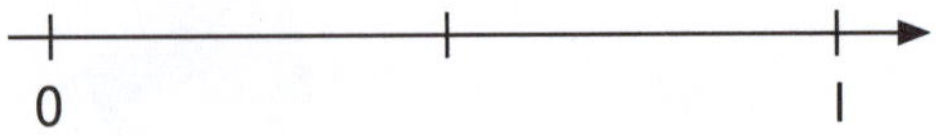

2. Start at 0 and count on two-thirds.

3. Start at 0 and count on three-quarters.

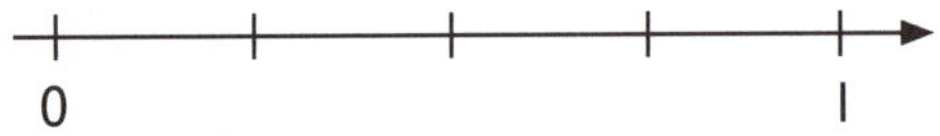

4. Start at 0 and count on three-fifths.

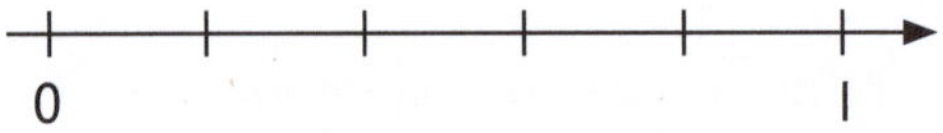

5. Start at 0 and count on three-sixths.

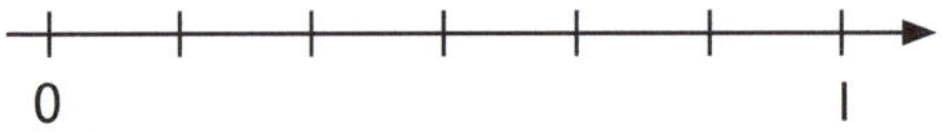

6. Start at 0 and count on five-eighths.

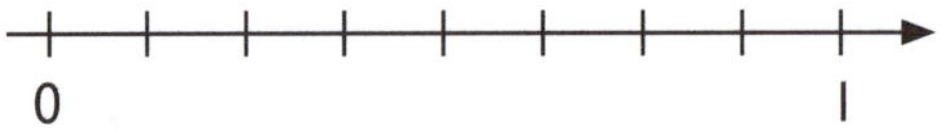

SCORE /6 0-2 3-4 5-6

Quick Check

Count on in fractions. Write the name of the fraction you end on in the box.

1.
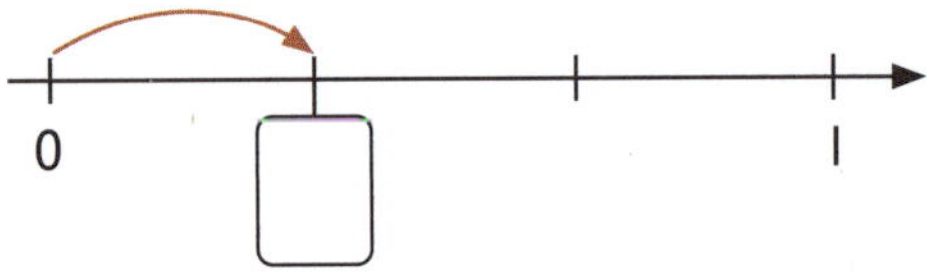

2.
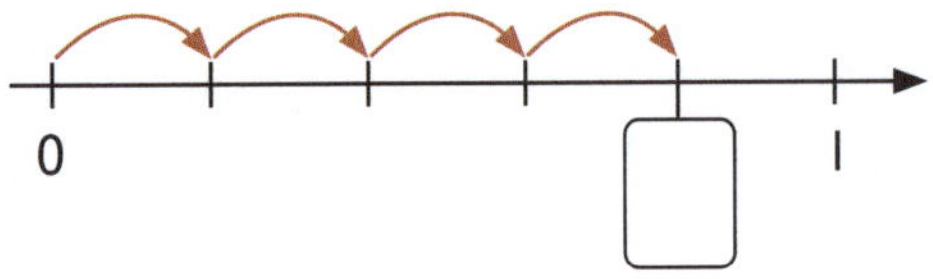

3.
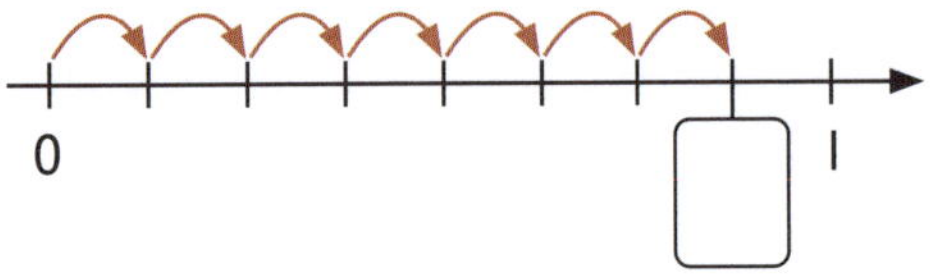

4.
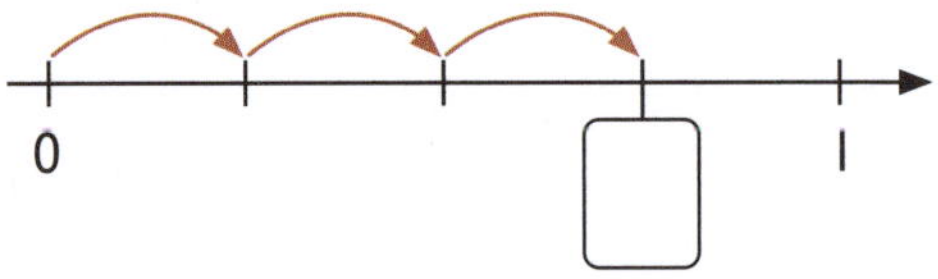

5.
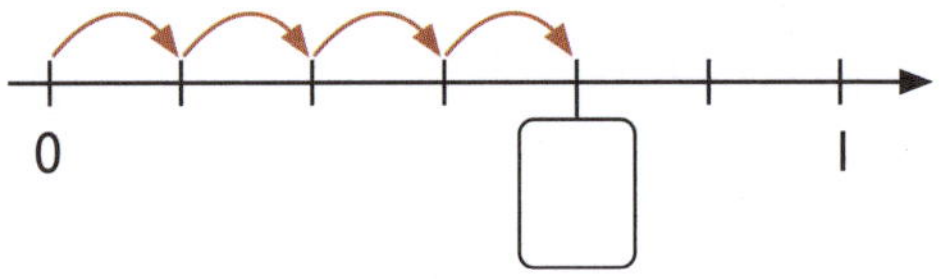

6.
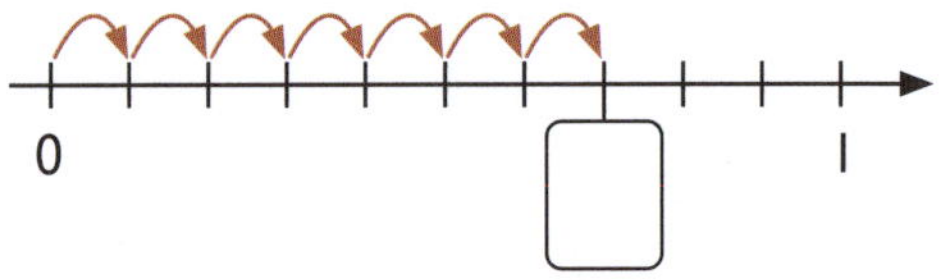

SCORE /6 0-2 3-4 5-6

TERM 4

TARGETING MATHS HOMEWORK: YEAR 3 © PASCAL PRESS ISBN: 9781925726565

Number & Algebra

AC9M3N03

Adding Strings of Numbers

To add 36, 14 and 25, add the tens first: 30 + 10 + 20 = 60

Then add the ones: 6 + 4 + 5 = 15

Add your results to get the sum of the three numbers: 60 + 15 = 75

Ada said, "I like to set it out like this:"

23 + 45 + 31

20 + 40 + 30 = 90

3 + 5 + 1 = 9

→ **99**

Use Ada's strategy for these additions.

1. 33 + 44 + 21
2. 47 + 22 + 13
3. 25 + 27 + 33
4. 28 + 43 + 26
5. 52 + 34 + 26
6. 72 + 31 + 47

SCORE /6 0-2 3-4 5-6

Statistics & Probability

AC9M3ST02

The Traffic Survey

Use the traffic survey results to answer these questions.

	7:00–8:00	8:00–9:00	9:00–10:00	10:00–11:00
Cars	卌 卌 |||	卌 卌 卌	卌 卌	卌 ||||
Utes	卌 卌 |||	卌 |||	卌	|||
Trucks	|||	||||	卌	卌 |
Buses	||||	卌 |	|||	|||
Taxis	|	||	|	|

1. At which time were the most utes on the road? ______________
2. Which was the busiest time? ______________
3. Why do you think that was the busiest time? ______________
4. Which vehicle was seen most often? ______________
5. How many vehicles were recorded altogether? _____

SCORE /5 0-2 3-4 5

AC9M3SP01

Stacking Blocks

You can stack and join blocks to build models.
How many blocks were used to build these models?

① ____ blocks

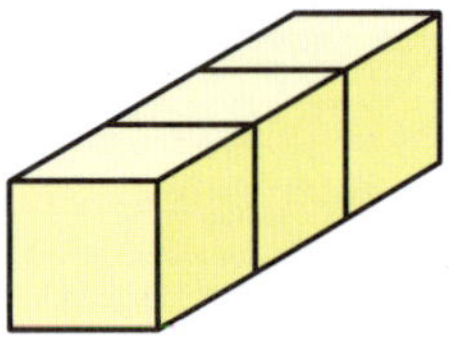

② ____ blocks

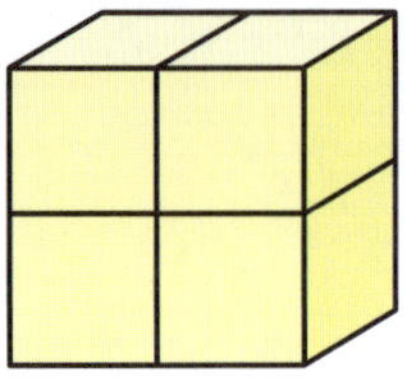

③ ____ blocks

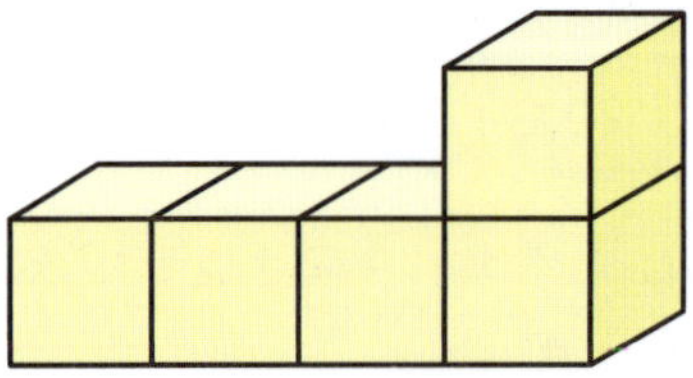

④ ____ blocks

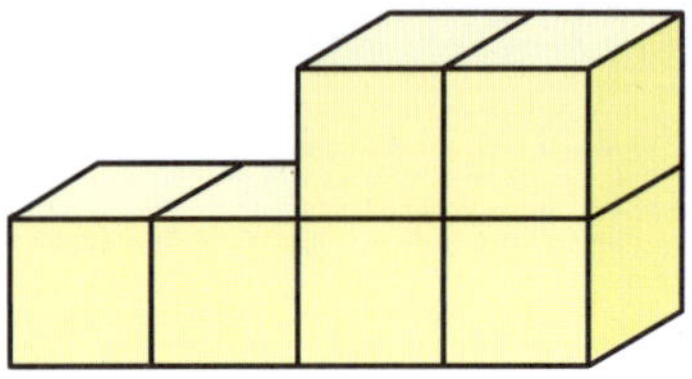

⑤ ____ blocks

⑥ ____ blocks

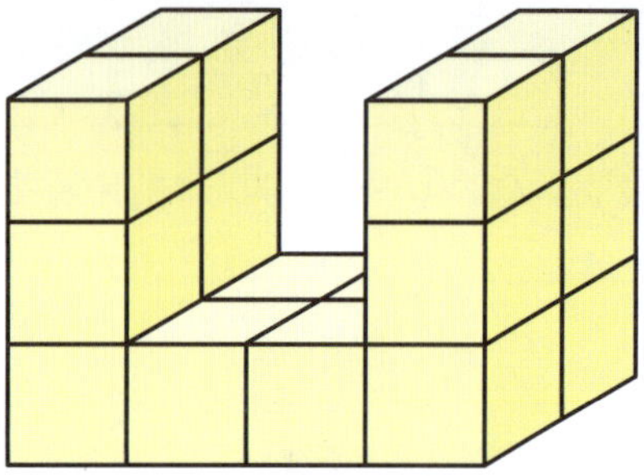

SCORE /6 0-2 3-4 5-6

NAPLAN*-style Questions

① What fraction is marked by ? on this number line?

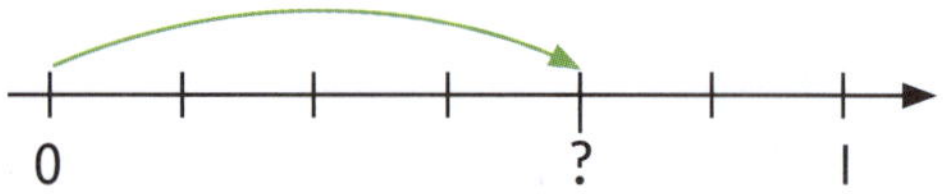

⬭ $\frac{1}{2}$ ⬭ $\frac{5}{4}$ ⬭ $\frac{4}{6}$ ⬭ $\frac{6}{9}$

② Bindi scored 26, 18 and 36 in the trivia quiz.
What was her total score?

☐

③ This model was made with cubes:

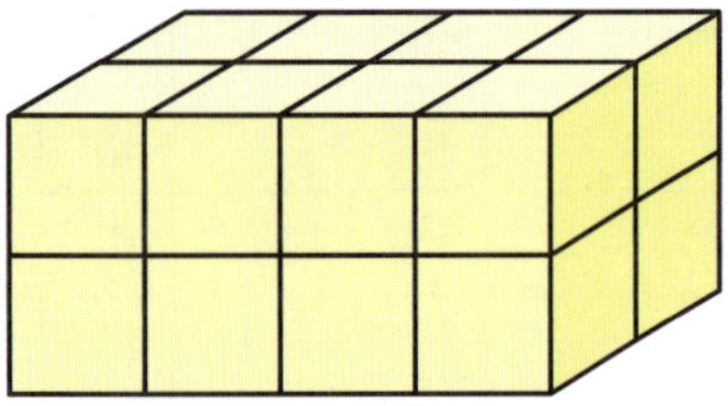

This is what the model looked like after some of the cubes were removed:

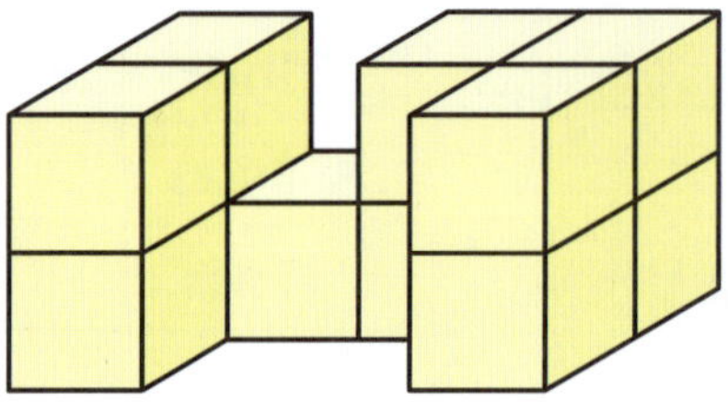

How many cubes were left in the model?

☐

SCORE /3 0-1 2 3

TARGETING MATHS HOMEWORK: YEAR 3 © PASCAL PRESS ISBN: 9781925726565

Work Smart, not Hard

"Work out 36 + 24 + 53 + 56," said Mr Green to his class.

"But I want you to **work smart, not hard** and show me your strategy for finding the answer," he added.

Here are the strategies that three students used.

This is Luis's strategy.

50 < 36, 24 > 10

100 < 53, 56 > 9

150 + 19 = 169

This is Max's strategy.

36 + 24 = 60

53 + 56 = 109

60, 109 > 169

This is Kiku's strategy.

36	+	24	+	53	+	56			
30	+	20	+	50	+	50	=	150	> 169
6	+	4	+	3	+	6	=	19	

Show how Luis, Max and Kiku could use their strategies to find 27 + 58 + 32 + 53.

Luis's strategy	**Max's strategy**
Kiku's strategy	**Try a strategy of you own.**

① Which strategy was easiest for you? ____________________

② Why was that? ____

TARGETING MATHS HOMEWORK: YEAR 3 © PASCAL PRESS ISBN: 9781925726565

Multiplication Turn-around Facts

Learning the multiplication tables is not as difficult as it might seem if you use what you already know.

If you are asked this question from the 7 times table: **What is 7 × 3?**

and don't know the answer, you might think that you don't know your 7 times table.
In fact, you do know that fact because you know that 3 × 7 = 21 so you also know that 7 × 3 = 21.

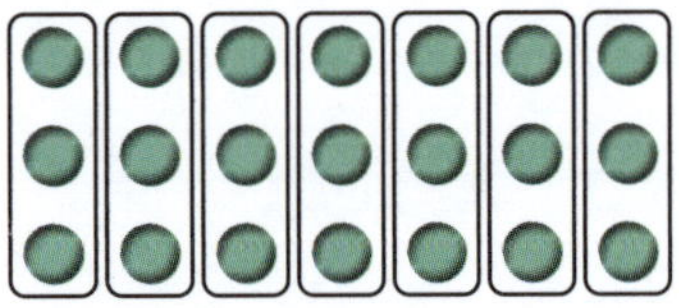

7 × 3 is the same as

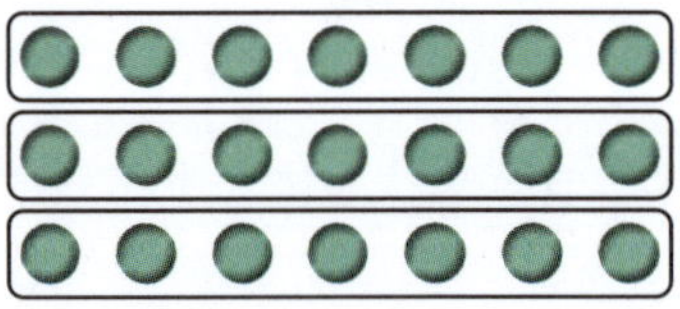

3 × 7

Multiplications like 7 × 3 and 3 × 7 are called **turn-arounds**.

Complete these turn-around facts.

1	2 × 6 = ☐	6 × 2 = ☐	8	9 × 3 = ☐	3 × 9 = ☐
2	7 × 5 = ☐	5 × 7 = ☐	9	2 × 12 = ☐	12 × 2 = ☐
3	5 × 8 = ☐	8 × 5 = ☐	10	9 × 4 = ☐	4 × 9 = ☐
4	8 × 4 = ☐	4 × 8 = ☐	11	7 × 4 = ☐	4 × 7 = ☐
5	6 × 3 = ☐	3 × 6 = ☐	12	5 × 11 = ☐	11 × 5 = ☐
6	3 × 8 = ☐	8 × 3 = ☐	13	7 × 8 = ☐	8 × 7 = ☐
7	6 × 4 = ☐	4 × 6 = ☐	14	8 × 9 = ☐	9 × 8 = ☐

SCORE /14 0-6 7-12 13-14

TERM 4

Quick Check

It is good to be really fast with your multiplication facts. If there is a fact that you don't know really quickly, put a circle around it to show you which facts to practise.

1. 2 × 6 = ☐
2. 5 × 5 = ☐
3. 6 × 3 = ☐
4. 10 × 5 = ☐
5. 7 × 5 = ☐
6. 8 × 2 = ☐
7. 7 × 4 = ☐
8. 8 × 5 = ☐
9. 9 × 2 = ☐
10. 10 × 6 = ☐
11. 7 × 10 = ☐
12. 2 × 5 = ☐
13. 9 × 4 = ☐
14. 9 × 5 = ☐
15. 10 × 10 = ☐
16. 2 × 8 = ☐
17. 3 × 7 = ☐
18. 5 × 8 = ☐
19. 10 × 9 = ☐
20. 4 × 4 = ☐

SCORE /20 0-9 10-18 19-20

TARGETING MATHS HOMEWORK: YEAR 3 © PASCAL PRESS ISBN: 9781925726565

Number & Algebra

UNIT 29

AC9M3N04, AC9M3A03

Multiplication and Division Fact Families

This is a multiplication and division fact family:

7 × 4 → 28   4 × 7 → 28   28 ÷ 7 → 4   28 ÷ 4 → 7

Use Ada's fact family to answer these word problems.

1. Ada has 28 balloons to make into 4 equal bunches.
   How many balloons should be in each bunch? ☐
2. A grandfather has $28 to share equally between his 7 grandchildren.
   How much should each grandchild get? ☐
3. Using the oranges I picked, I made 4 nets of oranges with 7 oranges in each net.
   How many oranges did I pick? ☐
4. 7 students each have a skateboard.
   How many wheels altogether? ☐

Complete this fact family for 18. Then write each fact beside the matching word problem.

3 × 6 → 18   ☐ × ☐ → ☐   18 ÷ ☐ → ☐   ☐ ÷ ☐ → ☐

5. 6 bags of onions with 3 onions in each bag.
   How many onions? ☐
6. 18 flowers to make 6 identical bunches.
   How many flowers? ☐
7. 3 packets with 6 pens in each packet.
   How many pens? ☐
8. 18 apples put into 3 bags.
   How many apples? ☐

SCORE /8 0-3 4-6 7-8

Statistics & Probability

AC9M3P02

A Chance Experiment

1. Make 6 identical small pieces of paper.
   Write the letter A on 3 pieces of paper, the letter B on 2 pieces of paper and the letter C on the last piece of paper. Put the pieces of paper into a bag that you can't see through.
   You are now ready to take one piece of paper out of the bag, record the letter written on it, put it back in the bag and give the bag a big shake.
2. If you do this 24 times, how many times do you think you will see each letter?
   A ____ times, B ____ times, C ____ times
3. Do the experiment 24 times. Record the results in this tally chart.
4. Did you get the result that you expected? ◯ Yes ◯ No
5. Why was that?

____

A	
B	
C	

SCORE /5 0-2 3-4 5

TERM 4

Fractions of a Measurement

Use the table of values to help you answer the following measurement questions.

Mass	Capacity	Length
1 kg = 1000 g	1 L = 1000 mL	1 m = 100 cm
$\frac{1}{2}$ kg = 500 g	$\frac{1}{2}$ L = 500 mL	$\frac{1}{2}$ m = 50 cm
$\frac{1}{4}$ kg = 250 g	$\frac{1}{4}$ L = 250 mL	$\frac{1}{4}$ m = 25 cm

Join the measurement fractions with the matching amounts.

1. $1\frac{1}{2}$ kg
2. $\frac{3}{4}$ L
3. $1\frac{1}{2}$ m
4. $1\frac{1}{4}$ L
5. $2\frac{1}{2}$ kg
6. $1\frac{1}{4}$ m

- 750 mL
- 1250 g
- 2500 g
- 1250 mL
- 125 cm
- 1500 g
- 150 cm

7. What measurement is left over? _______

SCORE /7 0-3 4-5 6-7

NAPLAN*-style Questions

1. Five children wanted to play marbles.
There were twenty-five marbles to share between them.
How many marbles should each child get?

☐

2. Look at Bindi's three bags of shape cards.
If Bindi chooses one card from each bag without looking, which combination of cards could she have chosen?

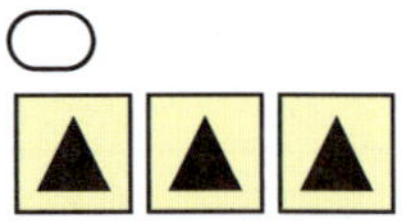

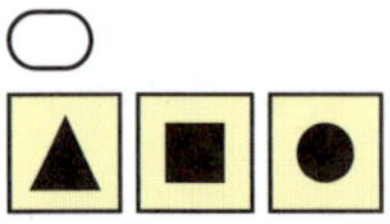

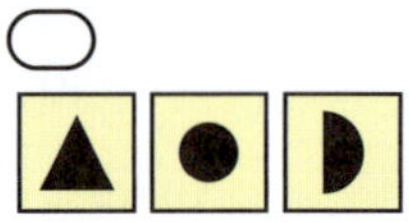

3. Luis bought six coloured markers at $4 each.
How much did the markers cost altogether?

$ ______

SCORE /3 0-1 2 3

TERM 4

TARGETING MATHS HOMEWORK: YEAR 3 © PASCAL PRESS ISBN: 9781925726565

AC9M3N06

Mamma Bella's Cafe

Bella, we have orders for:
- 6 pasta salads at tables 3, 4 and 7
- 5 burgers at tables 1, 2 and 8
- 4 pizzas at tables 5, 6, 9 and 10.

I need a list!
I'm too busy to work out how many of each I need to make.

Soon the waiter was back with more orders.

I have orders for:
- 5 ice-cream sundaes at tables 3, 4 and 7
- 6 fruit salads at tables 1, 2 and 8
- 4 apple crumbles at tables 5, 6, 9 and 10.

Make lists for Bella showing how many of each item she needs to make.

Bella's Mains List	Bella's Dessert List
Pasta salads: _____	Ice-cream sundaes: _____
Burgers: _____	Fruit salads: _____
Pizzas: _____	Apple crumbles: _____

With $10 for each main and $5 for each dessert, I wonder how much money we will make.

How much money will Mamma Bella's Cafe make today?

Mamma Bella's Cafe Takings

_____ @ $10 = _____

_____ @ $5 = _____

Total = $_______

UNIT 30

Number & Algebra

AC9M3A03

Rules for Multiplication

If you don't instantly know 3 × 3, there is a rule for the 3 times table that can help:

- Double the number to be multiplied and add one more of that number.

So for 3 × 2, double 2 to make 4 and add another 2 to get 6.

Use the rule for these multiplications.

1. 3 × 4
   2 × ☐ + ☐ = ☐
2. 3 × 5
   2 × ☐ + ☐ = ☐
3. 3 × 6
   2 × ☐ + ☐ = ☐
4. 3 × 7
   2 × ☐ + ☐ = ☐
5. 3 × 8
   2 × ☐ + ☐ = ☐
6. 3 × 9
   2 × ☐ + ☐ = ☐

**For 4 × 3, we know that 4 is made with 2 + 2, so we can use double-double.
Use the double-double rule for these multiplications.**

7. 4 × 6
   2 × ☐ + 2 × ☐ = ☐
8. 4 × 7
   2 × ☐ + 2 × ☐ = ☐
9. 4 × 8
   2 × ☐ + 2 × ☐ = ☐
10. 4 × 9
    2 × ☐ + 2 × ☐ = ☐

**For 6 × 3, we know that 6 is made with 5 + 1, and we know our 5 times table.
Use the × 5 + × 1 rule for these multiplications.**

11. 6 × 3
    5 × ☐ + 1 × ☐ = ☐
12. 6 × 7
    5 × ☐ + 1 × ☐ = ☐
13. 6 × 8
    5 × ☐ + 1 × ☐ = ☐
14. 6 × 9
    5 × ☐ + 1 × ☐ = ☐

SCORE /14 0-6 7-12 13-14

TERM 4

Quick Check

It is good to be really fast with your multiplication facts. If there is a fact that you don't know really quickly, put a circle around it to show you which facts to practise.

1. 4 × 1 = ☐
2. 3 × 5 = ☐
3. 4 × 9 = ☐
4. 5 × 10 = ☐
5. 6 × 2 = ☐
6. 3 × 8 = ☐
7. 9 × 3 = ☐
8. 6 × 3 = ☐
9. 10 × 7 = ☐
10. 9 × 2 = ☐
11. 6 × 4 = ☐
12. 3 × 6 = ☐
13. 3 × 11 = ☐
14. 4 × 11 = ☐

SCORE /14 0-6 7-12 13-14

TARGETING MATHS HOMEWORK: YEAR 3 © PASCAL PRESS ISBN: 9781925726565

Number & Algebra

AC9M3A02

Addition and Subtraction

We don't always have paper and pencil when we need to add or subtract, so we need some quick mental strategies to use.
This number line shows a mental strategy for 46 + 22, where 46 was left as is and 22 was split into 20 + 2.

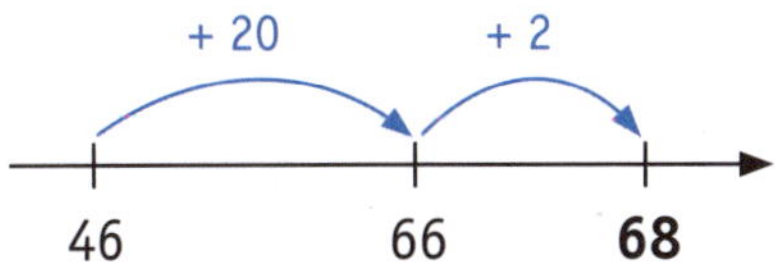

Use the strategy for these additions.

① 53 + 36

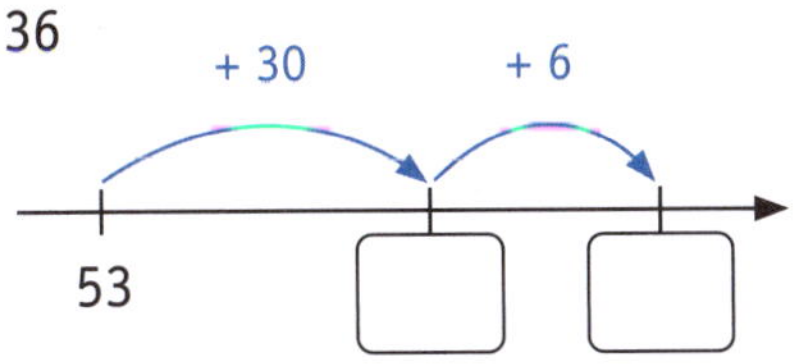

② 44 + 33

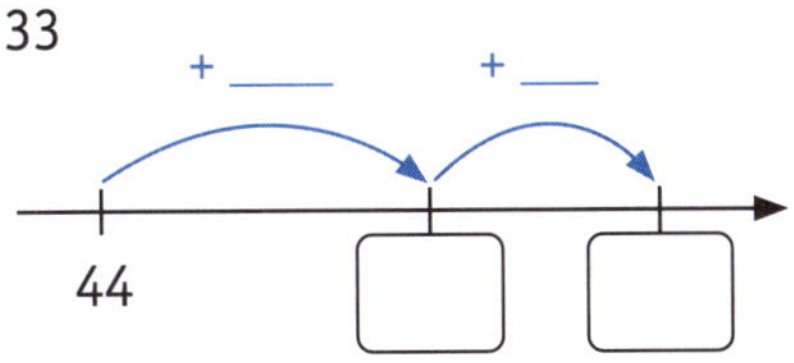

③ 54 + 25

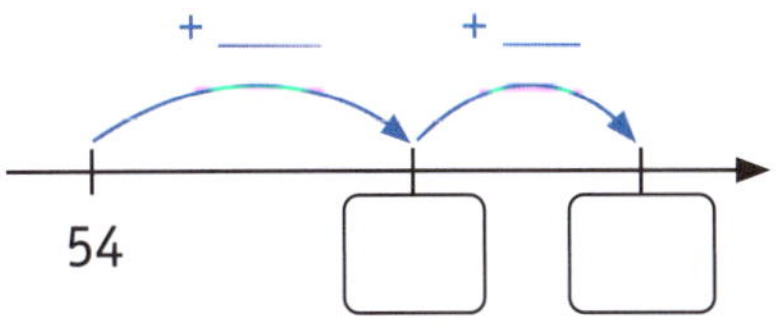

④ 38 + 23

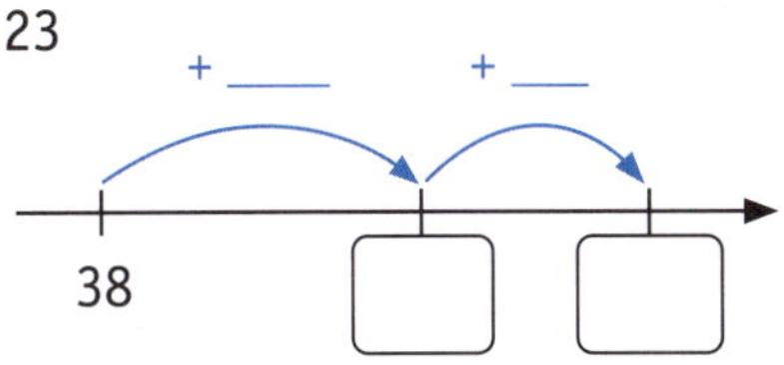

Now use the strategy for these subtractions.

⑤ 76 − 24

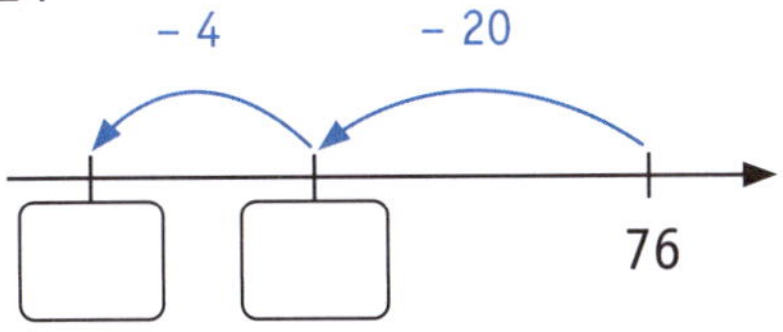

⑥ 49 − 31

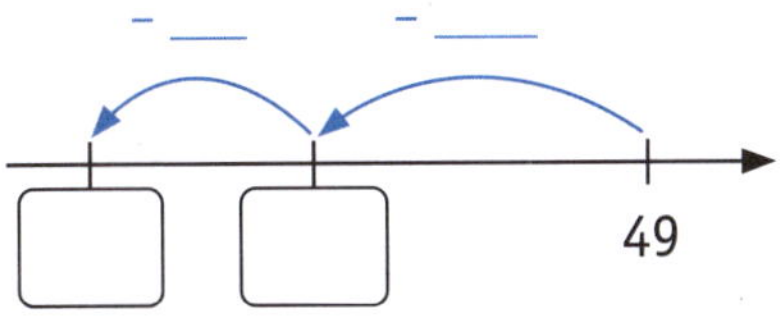

⑦ 83 − 51

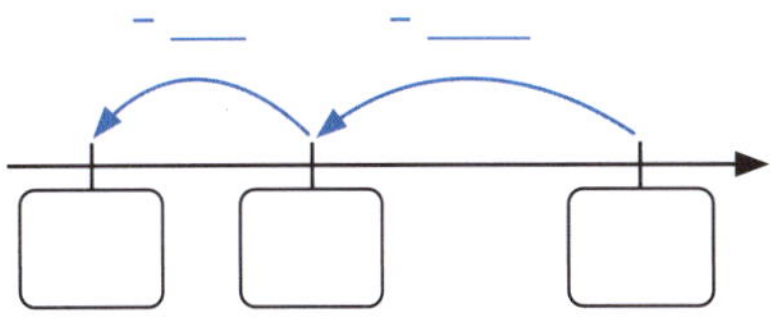

⑧ 108 − 82

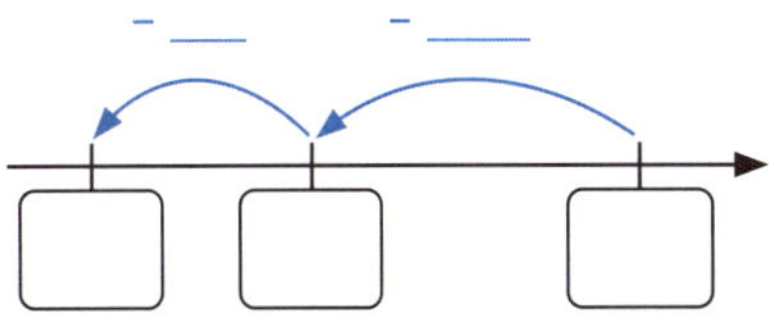

SCORE /8 0-3 4-6 7-8

Statistics & Probability

AC9M3P02

Probability

① On a scale of 1 to 3 where 1 = unlikely, 2 = probable and 3 = certain, how likely is it that you could talk to someone for one minute without saying "Um"?
Circle your answer on the number line.

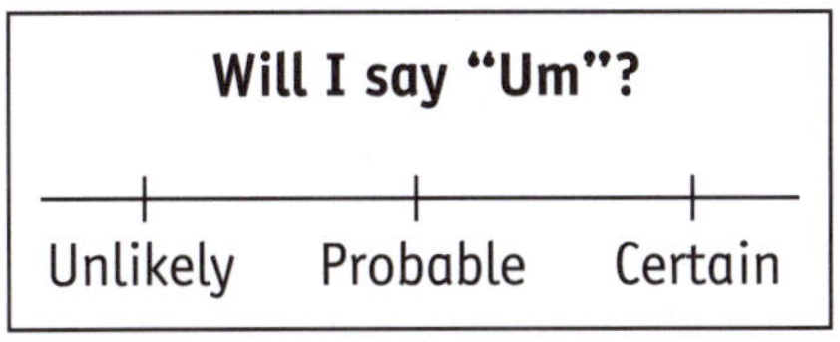

② Now set a timer and try it. What did you find?

____

____

SCORE /2 0 1 2

TERM 4

Measurement & Space

AC9M3M04, AC9M3M05

The Hands on the Clock

The hands on a clockface make angles.
At 11:30, the angle between the minute hand and the hour hand is **larger than** a right angle.

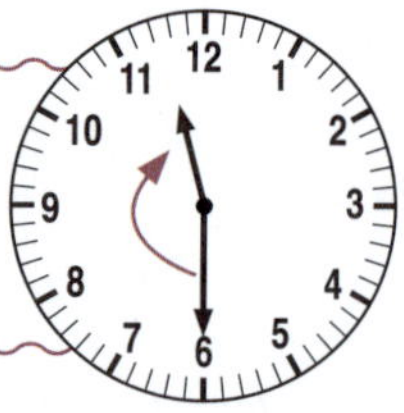

Show the times on the clocks.
Then write the missing words to describe the angle the hands make as larger than a right angle, close to a right angle or less than a right angle.

1. **3 o'clock**
   The angle the hands make is ____________________ a right angle.

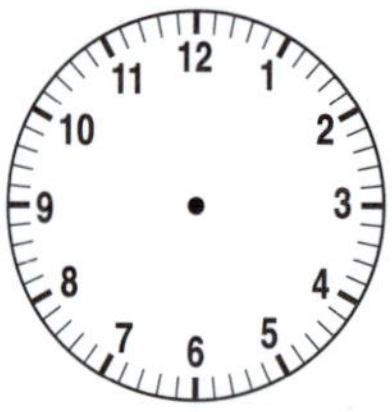

2. **7:30**
   The angle the hands make is ____________________ a right angle.

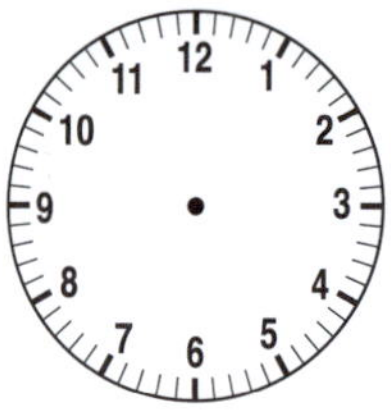

3. **Quarter past six**
   The angle the hands make is ____________________ a right angle.

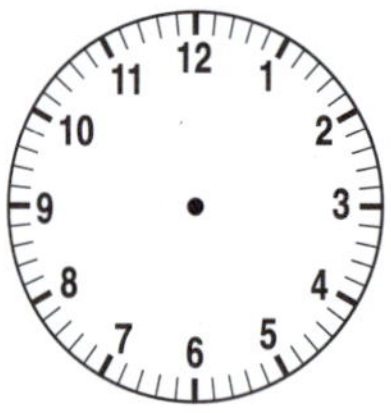

4. **Ten to three**
   The angle the hands make is ____________________ a right angle.

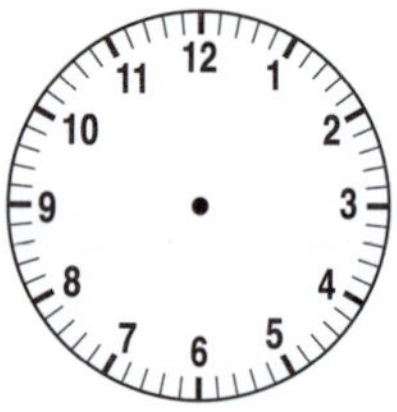

5. **6:45**
   The angle the hands make is ____________________ a right angle.

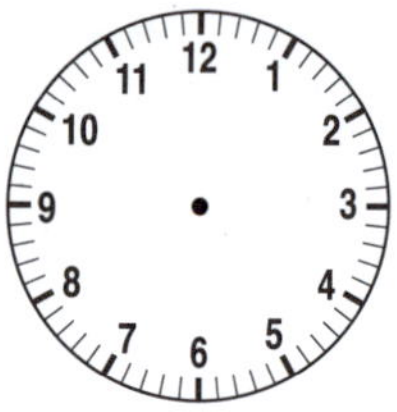

SCORE /10 (0-4) (5-8) (9-10)

NAPLAN*-style Questions

1. Kiku had $73. Then her aunt gave her $35 for her birthday.
   How much money does Kiku have now?
   $ [ ]

2. Hugo had $96. Then he spent $54.
   How much money does Hugo have now?
   $ [ ]

3. What time does this digital clock show?

   - ◯ Quarter past six
   - ◯ Quarter to five
   - ◯ Quarter to six
   - ◯ Quarter past five

SCORE /3 (0-1) (2) (3)

TERM 4

TARGETING MATHS HOMEWORK: YEAR 3 © PASCAL PRESS ISBN: 9781925726565

Making Totals

We have been making 2-digit numbers using only these digits:

1	2	3	4	5	6

We each made three numbers. Then we used smart strategies to add the numbers we made.

Here is an example:

12 + 34 + 56

10 + 30 + 50 = 90

2 + 4 + 6 = 12

90 and 12 → **102**

We are not going to tell you which numbers we each made but here are some clues so you can work it out.

① Ada: My total is 120.
Can you find a way to make that total?

____

③ Luis: I made the largest possible total.
What 2-digit numbers do you think I used?

____

② Bindi: I made the smallest possible total.
What 2-digit numbers do you think I used?

____

④ Max: The total of the numbers I made was 138.

____

We think there is more than one way of making our totals.
Now you have found one way, what other possibilities can you find?

⑤ Ada's total of 120 can also be made with

____

⑥ Bindi's total can also be made with

____

⑦ Luis's total can also be made with

____

⑧ Max's total can also be made with

____

TERM 4

Number & Algebra

AC9M3N01

Small and Large

Circle the smallest number in each group and write the largest number in words.

1. 50 653   63 504   7658

   Largest number: ____________________

2. 36 049   30 946   34 567

   Largest number: ____________________

3. 78 045   78 367   78 504

   Largest number: ____________________

4. 36 049   30 946   34 096

   Largest number: ____________________

5. 36 853   30 965   30 956

   Largest number: ____________________

6. 179 456   180 937   178 504

   Largest number: ____________________

7. 205 746   205 736   206 102

   Largest number: ____________________

In 76 576, there are 76 thousands.
How many thousands are in each of these numbers?

8. 36 457 ____ thousands
9. 6856 ____ thousands
10. 1247 ____ thousands
11. 123 489 ____ thousands
12. 10 659 ____ thousands
13. 847 ____ thousands

SCORE /13 0-6 7-11 12-13

Quick Check

Write the number that is 1 more.

1. 9999 ____________
2. 13 789 ____________
3. 99 999 ____________

Write the number that is 10 more.

4. 9999 ____________
5. 54 798 ____________
6. 99 999 ____________

Write the number that is 100 more.

7. 13 856 ____________
8. 24 956 ____________
9. 99 999 ____________

Write the number that is 1000 more.

10. 27 659 ____________
11. 29 656 ____________
12. 99 999 ____________

SCORE /12 0-5 6-10 11-12

TARGETING MATHS HOMEWORK: YEAR 3 © PASCAL PRESS ISBN: 9781925726565

Number & Algebra

AC9M3N06

Change from $100

You can use a number line to work out how much change to expect.

This number line shows the change you get when you spend $87 and pay with a $100 note:

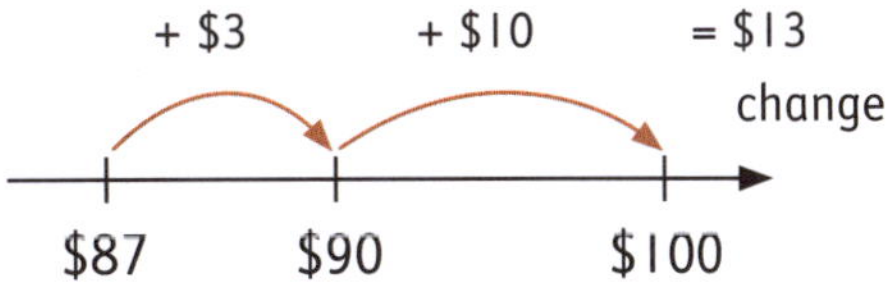

Use the number lines to work out the change from $100 for each of these purchases.

1. Buy a paint box for $36.

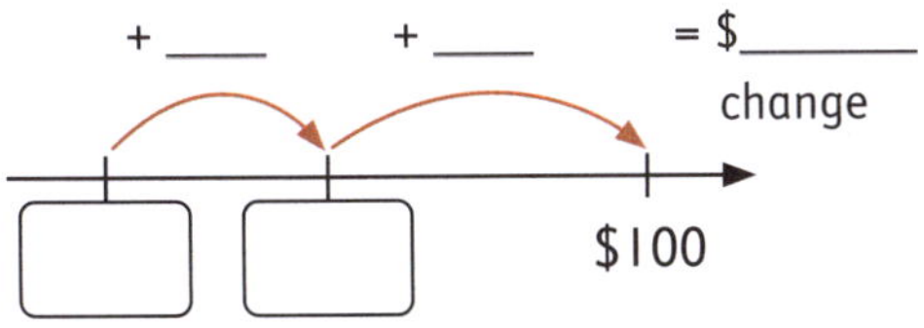

2. Buy two T-shirts for $42 each.

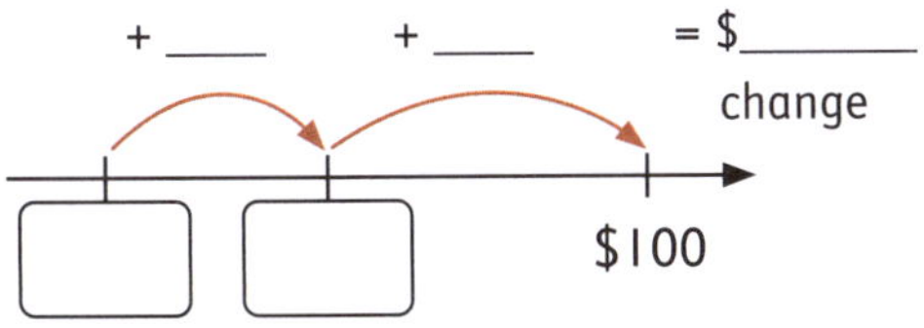

3. Buy 4 tacos at $9 each and 4 drinks at $4·50 each.

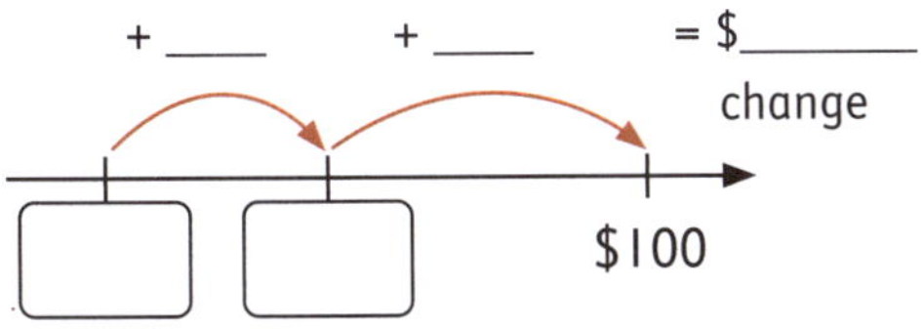

4. Buy an apple for $1·45.

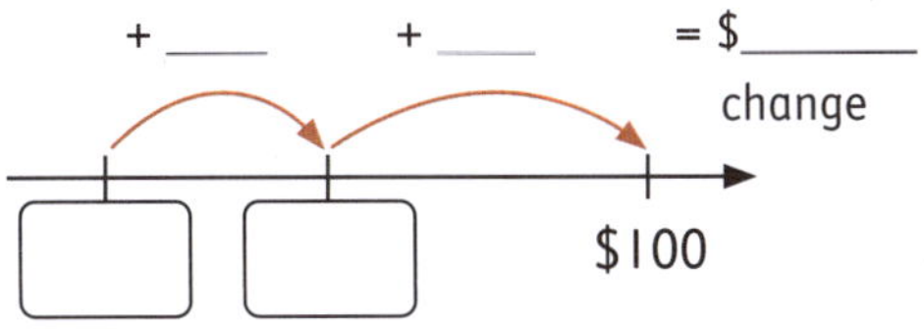

5. Buy a cap for $27.

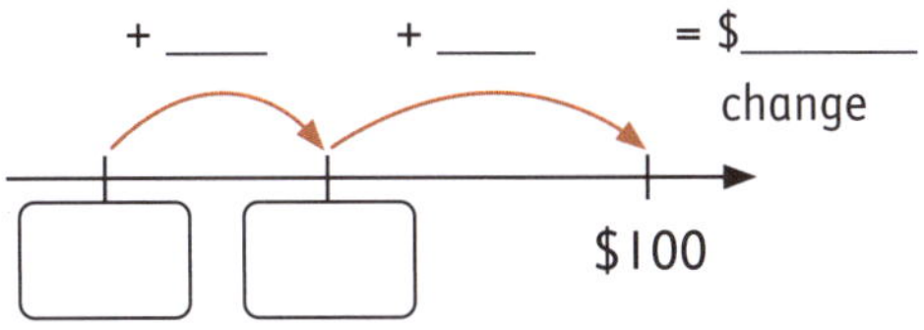

SCORE /5 0-2 3-4 5

Statistics & Probability

AC9M3ST02

The Lion Cubs

The zookeepers collect data on the growth of their lion cubs. They weigh each cub at the end of the week and record the weight. The table shows their data for Weeks 1 to 3.

Lion cub	Week 1	Week 2	Week 3
Eunice	1·5 kg	1800 g	2·5 kg
Vera	1400 g	1800 g	2300 g
Wallace	1200 g	1·5 kg	2200 g

1. Which cub was heaviest at the end of Week 1? ____________
2. Which cub gained the most weight during Week 2? ____________
3. Which cubs gained the most weight during Week 3? ____________
4. Which cubs gained the most weight between the end of Week 1 and the end of Week 3?

____

5. If Wallace gained another 700 g next week, how heavy would he be then? ____________

SCORE /5 0-2 3-4 5

TERM 4

Time Spans

What would be the approximate duration of these events?

① Hugo skipped 100 times. ____________

② Ada baked a cake. ____________

③ Bindi watched a movie. ____________

④ Luis counted to sixty. ____________

Now give approximate answers to these time questions.

⑤ How many days in one month? ____________

⑥ How many hours between midnight and noon? ____________

⑦ How many days between full moons? ____________

⑧ How many days from New Year's Day to New Year's Eve? ____________

⑨ How many months in a year? ____________

⑩ How many days in a season? ____________

SCORE /10 0-4 5-8 9-10

NAPLAN*-style Questions

① Max bought some crayons with a $10 note.
The crayons were $1·20 each and Max bought 6 of them.
How much change did he get?

◯ $4 ◯ $2·80 ◯ $8·80 ◯ $7·20

② What is the largest number that can be made using all of these digits with no repeats?

0 2 4 6 8 5

◯ 856 240 ◯ 802 645 ◯ 865 420 ◯ 685 042

③ Look at the graph of cat colours.

Which of these is a correct statement about the data?

◯ Ginger cats are the most common.

◯ There are 2 more tabby cats than there are ginger cats.

◯ There are 5 tabby cats.

◯ There are 6 black cats.

SCORE /3 0-1 2 3

TARGETING MATHS HOMEWORK: YEAR 3 © PASCAL PRESS ISBN: 9781925726565

Problem of the Week

AC9M3NO1

Fun with a Calculator

Kiku, Hugo and Ada like to enter a 6-digit number like 643 578 into a calculator and then eliminate one number at a time from the display.

To eliminate the 3 from this display, you have to enter – 3000.

If you entered – 3, that would leave 643 575 in the display.

1. You have 6 moves to eliminate the whole display.
Try this example and then check on your calculator.

Start 736 859

Move 1:	–		leaves	
Move 2:	–		leaves	
Move 3:	–		leaves	
Move 4:	–		leaves	
Move 5:	–		leaves	
Move 6:	–		leaves	

What if you have to use the plus key at least once in your moves?

For example, if your start number was 643 578, you could use + 500 to eliminate the 5 and that would leave 644 078.

2. Include at least one addition move to eliminate the number displayed on this calculator.

Start 856 795

Move 1:		leaves	
Move 2:		leaves	
Move 3:		leaves	
Move 4:		leaves	
Move 5:		leaves	
Move 6:		leaves	

Did you check on your calculator?

TERM 4

Number & Algebra

AC9M3N04, AC9M3N05

Estimation for Division

If you don't know 96 ÷ 4, you can use multiplication by 4 to get close to 96.
4 × 20 = 80, so we know that the result of 96 ÷ 4 will be close to but more than 20.
Let's check.
96 ÷ 4 = 24, so the estimate of 20 was quite close to the correct answer.

Estimate and then check the answers to these divisions.

	Division	A Near Multiplication	Estimate	Check
1	132 ÷ 6	6 × ☐ = 120	more than 20	132 ÷ 6 = ☐
2	84 ÷ 3	3 × ☐ = 90	less than 30	84 ÷ 3 = ☐
3	76 ÷ 4	4 × ☐ = ☐	____________	76 ÷ 4 = ☐
4	95 ÷ 5	5 × ☐ = ☐	____________	95 ÷ 5 = ☐
5	132 ÷ 4	4 × ☐ = ☐	____________	132 ÷ 4 = ☐
6	78 ÷ 6	6 × ☐ = ☐	____________	78 ÷ 6 = ☐
7	154 ÷ 7	7 × ☐ = ☐	____________	154 ÷ 7 = ☐
8	108 ÷ 9	9 × ☐ = ☐	____________	108 ÷ 9 = ☐
9	294 ÷ 3	3 × ☐ = ☐	____________	294 ÷ 3 = ☐
10	138 ÷ 6	6 × ☐ = ☐	____________	138 ÷ 6 = ☐

SCORE /10 0-4 5-8 9-10

Quick Check

Complete these related division and multiplication facts.

1. 24 ÷ 6 = ____
   4 × 6 = ____
2. 21 ÷ 7 = ____
   ____ × 7 = ____
3. 32 ÷ 4 = ____
   ____ × ____ = ____
4. ____ ÷ ____ = ____
   9 × 6 = 54
5. ____ ÷ ____ = ____
   5 × 7 = ____
6. 48 ÷ 4 = ____
   ____ × ____ = ____
7. ____ ÷ ____ = ____
   3 × 9 = 27
8. 36 ÷ 4 = ____
   ____ × ____ = ____
9. ____ ÷ ____ = ____
   5 × 9 = ____
10. 44 ÷ 4 = ____
    ____ × ____ = ____

SCORE /10 0-4 5-8 9-10

TARGETING MATHS HOMEWORK: YEAR 3 © PASCAL PRESS ISBN: 9781925726565

AC9M3N04, AC9M3N05

Estimation for Multiplication

Here is an estimation strategy for multiplication.

For 96 × 4, estimate with 100 × 4 = 400.

The estimate is 400.

Then use number splitting to check.

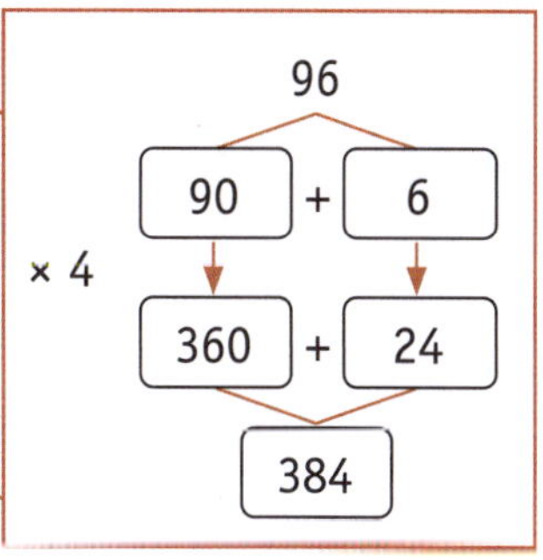

Use the estimate–check method with these examples.

① For 83 × 3, estimate with

☐ × ☐ = ☐

Check:

× 3

② For 72 × 4, estimate with

☐ × ☐ = ☐

Check:

× ___

③ For 36 × 4, estimate with

☐ × ☐ = ☐

Check:

× ___

④ For 23 × 6, estimate with

☐ × ☐ = ☐

Check:

× ___

⑤ For 39 × 3, estimate with

☐ × ☐ = ☐

Check:

× ___

⑥ For 106 × 4, estimate with

☐ × ☐ = ☐

Check:

× ___

SCORE /10 0-4 5-8 9-10

Statistics & Probability

AC9M3ST03

Frequently Used Letters

Max said that A was the most frequently used letter in writing, but Kiku said that it was O. Hugo was certain that it was E. Use the chart to make a tally chart of the number of times the letters A, O and E have been used in this paragraph.

Tally Chart of Letter Frequency

A	
O	
E	

① Do you agree with Max, Kiku or Hugo? ___________

② Why did you choose that person? ___________________________________

③ How do you think the frequencies of I and U compare with the frequencies of A, O and E?

____

SCORE /6 0-2 3-4 5-6

TARGETING MATHS HOMEWORK: YEAR 3 © PASCAL PRESS ISBN: 9781925726565

Giving Change

Change should always be given with the smallest number of coins and notes as possible. You can use a number line to work out which coins and notes to use.

I had $10 to spend and bought a box of cookies for $4·25.

This is how I checked that I had the right change with the smallest number of coins and notes:

5c 20c 50c $5 = $5·75 change

$4·25 $4·30 $4·50 $5 $10

Complete these examples of giving change for $10.

1. $5 = $_____ change

   $3·75 $3·80 $4·00 $5 $10

2. = $_____ change

   $2·80 $10

3. = $_____ change

   $4·10 $10

4. = $_____ change

   $6·55 $10

5. = $_____ change

   $3·05 $10

SCORE /5 0-2 3-4 5

NAPLAN*-style Questions

1. Ada baked 24 muffins. She put them in 4 equal rows on the muffin tray. How many muffins were in each row?

   ◯ 28 ◯ 20 ◯ 6 ◯ 4

2. Bindi wants to know how many oranges altogether would be in 6 bags of 8 oranges. Which number sentences match this problem?

   ◯ 6 + 8 = 14 ◯ 6 × 8 = 48 ◯ 8 × 6 = 48 ◯ 8 – 6 = 2

3. Luis has $7·20 in his money box. How much more does he need to make it up to $10?

   $ _____

SCORE /3 0-1 2 3

TERM 4

TARGETING MATHS HOMEWORK: YEAR 3 © PASCAL PRESS ISBN: 9781925726565

Multiplying Dice Game

Max and Kiku have been playing a multiplying dice game.

They each rolled two dice and then multiplied the two numbers to make their product.

The person with the highest product won that round.

As they played, they noticed that there were some products that could not be made and some that could be made in different ways. So they began an investigation to answer these questions:

- Which products can you make?
- Which products can be made in more than one way?
- Which products cannot be made at all?

3 × 4 = 12

Complete the investigation for Max and Kiku.

Products that can be made and ways of making them	Products that cannot be made
1 = 1 × 1	7
2 = 2 × 1	11
3 = ___ × ___	
4 = 4 × 1 or 2 × 2	

TERM 4 REVIEW

Review: Unit 25 and Unit 26

AC9M3N01, AC9M3N02, AC9M3N06, AC9M3N07, AC9M3M04, AC9M3ST02

Write these numbers in numerals.

1. Twenty-five thousand, six hundred and seventy-two ☐
2. Fifty thousand, four hundred and sixty-five ☐
3. Eighteen thousand and seventy-six ☐

How many thousands are there in these numbers?

4. 38 432 ☐ thousands
5. 20 639 ☐ thousands

Continue these number sequences.

6. 29, 31, 33, ____, ____, ____
7. 56, 54, 59, 57, ____, ____, ____

Use the timelines to work out how long between the start time and the finish time.

8. Start time 3:40 ———— 4:15 Finish time = ___ min
9. Start time 11:45 ———— 12:30 Finish time = ___ min

10. How much pizza is left if $\frac{1}{4}$ has been eaten? ___
11. How much chocolate would be left if $\frac{3}{8}$ has been eaten? ___

Complete the bar models to answer these questions.

12. Hugo has four more stickers than Ada. Together they have eighteen stickers.

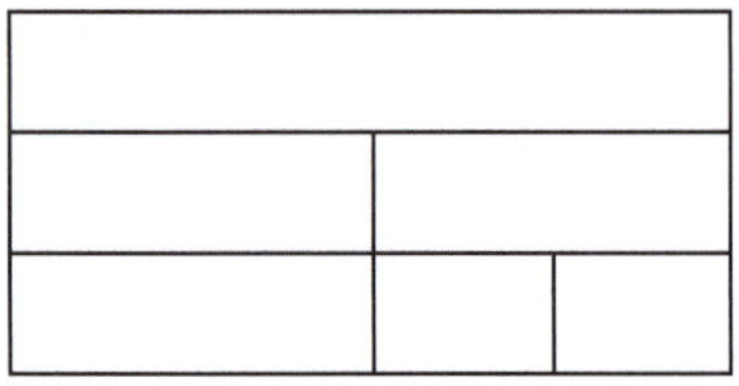

How many stickers does Hugo have? _____

13. Luis has six fewer strawberries than Bindi. Together they have twenty-eight strawberries. How many strawberries do they each have?

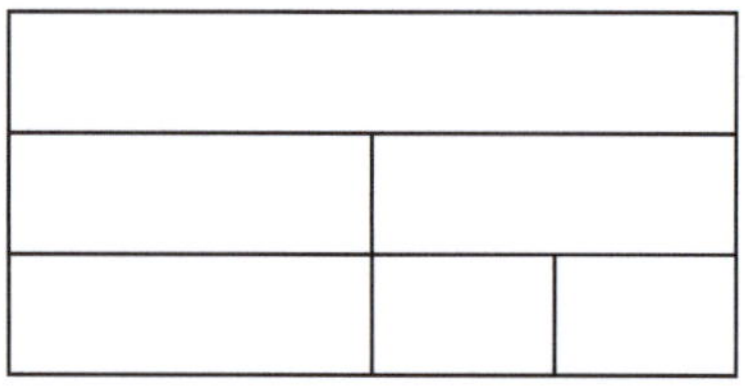

Luis has _____ strawberries.

Bindi has _____ strawberries.

14. Show the tally data on the bar graph.

Coloured Counters

Red						
Blue						
Yellow						

☐ = 2

Red counters: 𝍸 𝍸

Blue counters: 𝍸

Yellow counters: 𝍸 |||

SCORE /14 0-6 7-12 13-14

TARGETING MATHS HOMEWORK: YEAR 3 © PASCAL PRESS ISBN: 9781925726565

Review: Unit 27 and Unit 28

AC9M3N02, AC9M3N03, AC9M3N05, AC9M3N06, AC9M3SP01

1. About how many counters are hidden in this picture? Estimate then count to check.

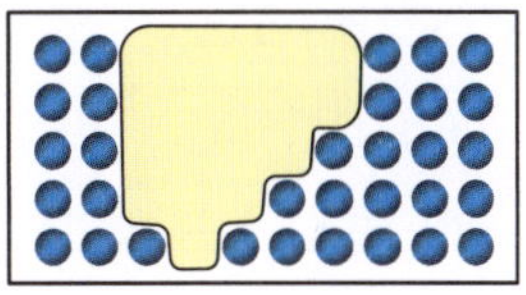

Estimate: ____

Actual: ____

Estimate these amounts to the nearest $10.

2. $756 $______
3. $704 $______
4. $897 $______
5. $996 $______
6. $1399 $______

Write the number sentences and work out the answers.

7. Bindi has five tubes of tennis balls with eight balls in each tube. How many tennis balls does Bindi have?

____

8. Luis started with $100 and spent $64. How much money does Luis have now?

____

Count on and write the fraction you end on in number form.

9. Start at 0 and count on three-eighths.

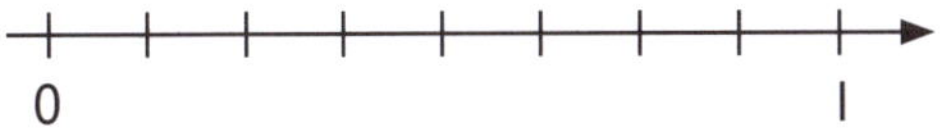

10. Start at 0 and count on two-fifths.

Add the tens and then the ones. Then add the results.

11. 43 + 25 + 31

☐ + ☐ + ☐ = ☐
☐ + ☐ + ☐ = ☐ → ☐

12. 33 + 44 + 32

☐ + ☐ + ☐ = ☐
☐ + ☐ + ☐ = ☐ → ☐

How many blocks were used to build these models?

13. ____ blocks

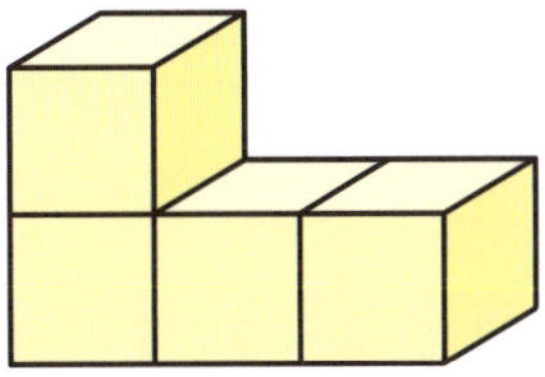

14. ____ blocks

15. ____ blocks

16. ____ blocks

SCORE /16 0-7 8-14 15-16

Review: Unit 29 and Unit 30

AC9M3N04, AC9M3A02, AC9M3A03, AC9M3M01, AC9M3M04, AC9M3M05

Complete these turn-around facts.

① 3 × 6 = ☐ 6 × 3 = ☐

② 7 × 4 = ☐ ☐ × ☐ = ☐

③ 4 × 9 = ☐ ☐ × ☐ = ☐

④ 3 × 7 = ☐ ☐ × ☐ = ☐

Write the number sentences and work out the answers.

⑤ There are 4 bags with 9 onions in each bag.
How many onions altogether?

4 × ☐ = ☐ onions

⑥ 24 salami slices are shared equally among 6 pizzas.
How many salami slices are on each pizza?

☐ ÷ ☐ = ☐ slices

⑦ How many grams are there in $1\frac{1}{2}$ kilograms? ________ g

⑧ How many centimetres are there in $1\frac{1}{2}$ metres? ________ cm

⑨ How many millilitres are there in $2\frac{1}{2}$ litres? ________ mL

Use the **multiply by 3** rule for these examples.
(Double the number to be multiplied and add one more of that number.)

⑩ 3 × 6 = ☐ × ☐ + ☐ = ☐

⑪ 3 × 9 = ☐ × ☐ + ☐ = ☐

Use the **double–double** strategy to multiply these numbers by 4.

⑫ 4 × 7 = ☐ × ☐ + ☐ × ☐ = ☐

⑬ 4 × 9 = ☐ × ☐ + ☐ × ☐ = ☐

Use the number lines to complete these additions and subtractions.

⑭ 44 + 53

+ ____ + ____

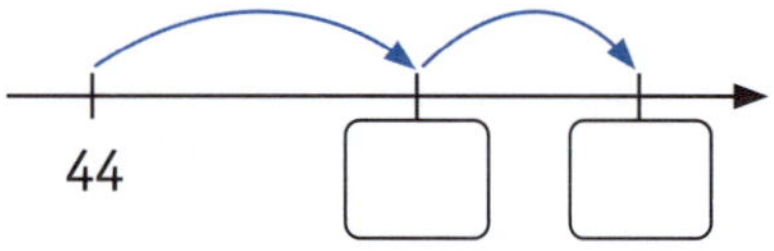

⑯ 87 – 53

– ____ – ____

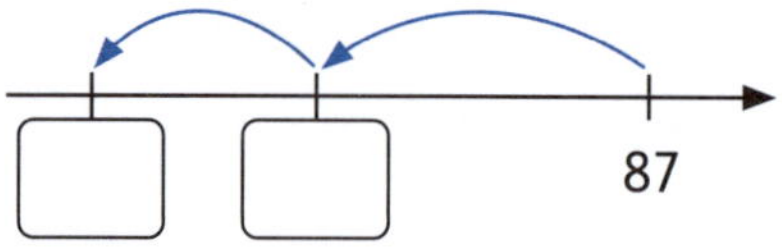

⑮ 62 + 48

+ ____ + ____

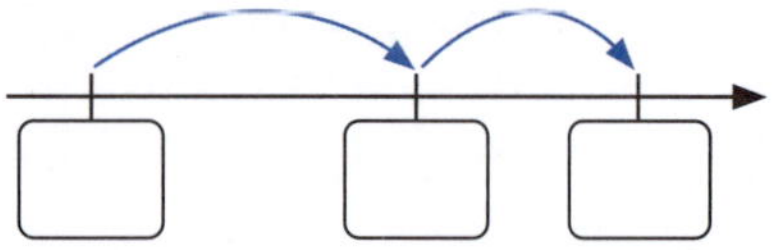

⑰ 96 – 57

– ____ – ____

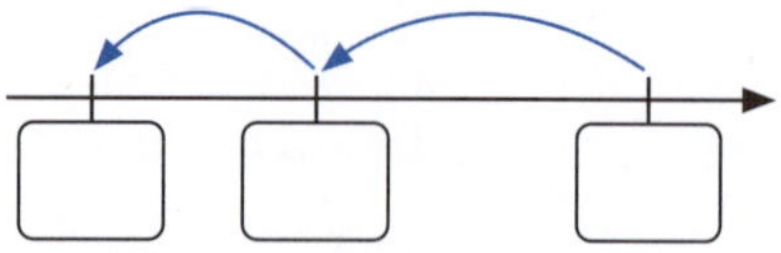

What times do these clocks show? Is the angle between the hands a right angle?

⑱ Time: ________

Right angle?

◯ Yes

◯ No

⑲ Time: ________

Right angle?

◯ Yes

◯ No

SCORE /19 0-9 10-17 18-19

TERM 4

TARGETING MATHS HOMEWORK: YEAR 3 © PASCAL PRESS ISBN: 9781925726565

Review: Unit 31 and Unit 32

AC9M3N01, AC9M3N04, AC9M3N05, AC9M3N06, AC9M3A03, AC9M3ST02

Write these numbers in numerals.

1. Sixty-five thousand, four hundred and seventy-two ____________
2. Thirteen thousand and sixty four. ____________

Add 100 to each of these numbers.

3. 7659 __________
4. 13 758 __________
5. 17 956 __________

Use the number lines to work out the change from $100.

6. Spend $27.

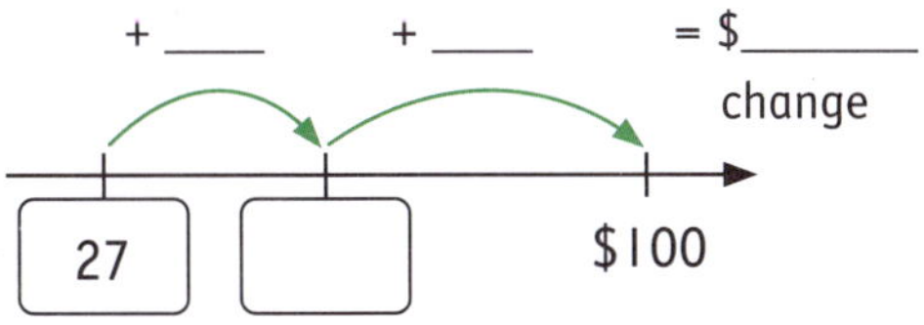

7. Spend $43.

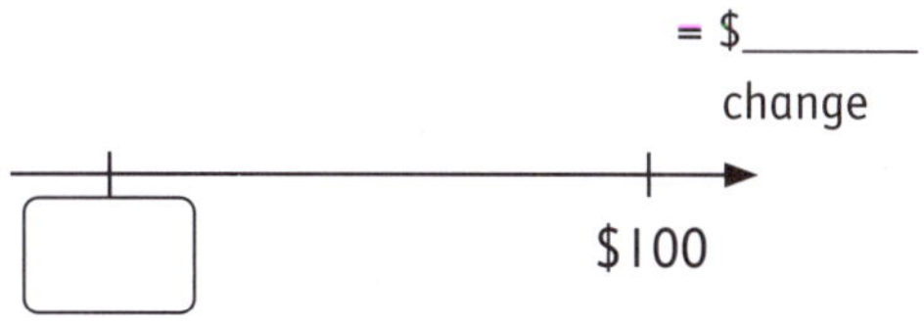

Look at the table.

How much did each lion cub grow between the end of Week 1 and the end of Week 2?

8. Eunice __________
9. Vera __________
10. Wallace __________

The Lion Cubs' Growth

Lion cub	Week 1	Week 2
Eunice	2·5 kg	3 kg
Vera	2300 g	3 kg
Wallace	2200 g	2·5 kg

Estimate the answers to these divisions.

11. 96 ÷ 4

    Estimate: ____________

12. 56 ÷ 4

    Estimate: ____________

13. 72 ÷ 6

    Estimate: ____________

Complete these related multiplication and division facts.

14. 27 ÷ 3 = ☐

    ☐ × ☐ = ☐

15. 36 ÷ 4 = ☐

    ☐ × ☐ = ☐

Use the estimate and check method for these multiplications.

16. For 54 × 6, estimate with

    ☐ × ☐ = ☐

    Check:

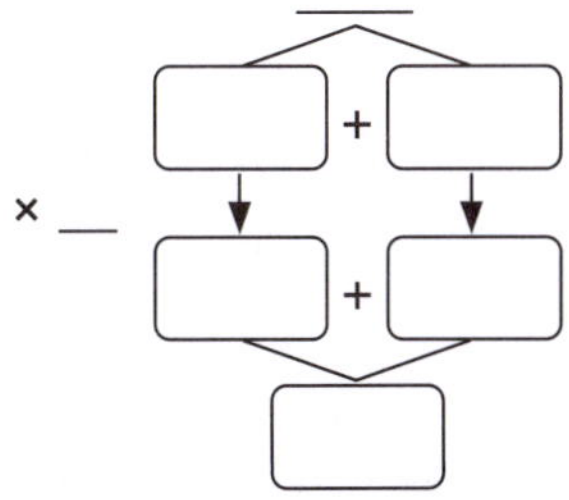

17. For 39 × 5, estimate with

    ☐ × ☐ = ☐

    Check:

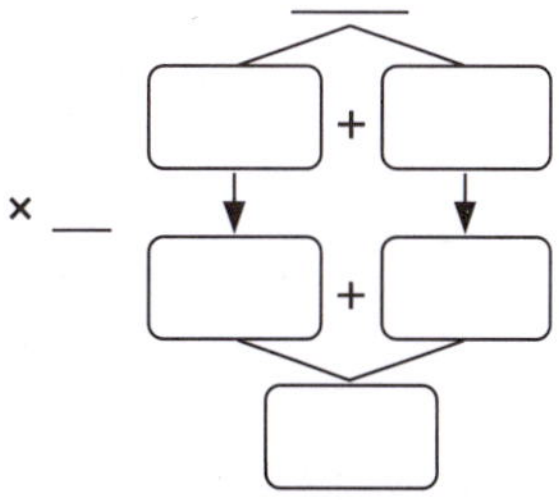

SCORE /17 0-8 9-15 16-17

★ QUICK MATHS FACTS ★

Measurement

Length

1 km = 1000 m
1 m = 100 cm
1 cm = 10 mm

Area

1 m^2 = 10 000 cm^2
1 hectare = 10 000 m^2
1 km^2 = 1 000 000 m^2

Capacity

1 L = 1000 mL

Mass

1 t = 1000 kg
1 kg = 1000 g

Fractions, decimals and percentages

$\frac{1}{100}$ = 0·01 = 1%

$\frac{25}{100}$ = 0·25 = 25%

$\frac{50}{100}$ = 0·5 = 50%

$\frac{75}{100}$ = 0·75 = 75%

$\frac{100}{100}$ = 1 = 100%

Place value

Ones	1
Tens	12
Hundreds	123
Thousands	1234
Ten thousands	12 345
Hundred thousands	123 456
Millions	1 234 567
Ten millions	12 345 678
Hundred millions	123 456 789

Time

60 s = 1 min
60 min = 1 h
1440 min = 1 day
24 h = 1 day
7 days = 1 week
365 days = 1 year
52 weeks = 1 year
12 months = 1 year

Multiplication tables

1 × 1 = 1	2 × 1 = 2	3 × 1 = 3	4 × 1 = 4	5 × 1 = 5	6 × 1 = 6
1 × 2 = 2	2 × 2 = 4	3 × 2 = 6	4 × 2 = 8	5 × 2 = 10	6 × 2 = 12
1 × 3 = 3	2 × 3 = 6	3 × 3 = 9	4 × 3 = 12	5 × 3 = 15	6 × 3 = 18
1 × 4 = 4	2 × 4 = 8	3 × 4 = 12	4 × 4 = 16	5 × 4 = 20	6 × 4 = 24
1 × 5 = 5	2 × 5 = 10	3 × 5 = 15	4 × 5 = 20	5 × 5 = 25	6 × 5 = 30
1 × 6 = 6	2 × 6 = 12	3 × 6 = 18	4 × 6 = 24	5 × 6 = 30	6 × 6 = 36
1 × 7 = 7	2 × 7 = 14	3 × 7 = 21	4 × 7 = 28	5 × 7 = 35	6 × 7 = 42
1 × 8 = 8	2 × 8 = 16	3 × 8 = 24	4 × 8 = 32	5 × 8 = 40	6 × 8 = 48
1 × 9 = 9	2 × 9 = 18	3 × 9 = 27	4 × 9 = 36	5 × 9 = 45	6 × 9 = 54
1 × 10 = 10	2 × 10 = 20	3 × 10 = 30	4 × 10 = 40	5 × 10 = 50	6 × 10 = 60
1 × 11 = 11	2 × 11 = 22	3 × 11 = 33	4 × 11 = 44	5 × 11 = 55	6 × 11 = 66
1 × 12 = 12	2 × 12 = 24	3 × 12 = 36	4 × 12 = 48	5 × 12 = 60	6 × 12 = 72

7 × 1 = 7	8 × 1 = 8	9 × 1 = 9	10 × 1 = 10	11 × 1 = 11	12 × 1 = 12
7 × 2 = 14	8 × 2 = 16	9 × 2 = 18	10 × 2 = 20	11 × 2 = 22	12 × 2 = 24
7 × 3 = 21	8 × 3 = 24	9 × 3 = 27	10 × 3 = 30	11 × 3 = 33	12 × 3 = 36
7 × 4 = 28	8 × 4 = 32	9 × 4 = 36	10 × 4 = 40	11 × 4 = 44	12 × 4 = 48
7 × 5 = 35	8 × 5 = 40	9 × 5 = 45	10 × 5 = 50	11 × 5 = 55	12 × 5 = 60
7 × 6 = 42	8 × 6 = 48	9 × 6 = 54	10 × 6 = 60	11 × 6 = 66	12 × 6 = 72
7 × 7 = 49	8 × 7 = 56	9 × 7 = 63	10 × 7 = 70	11 × 7 = 77	12 × 7 = 84
7 × 8 = 56	8 × 8 = 64	9 × 8 = 72	10 × 8 = 80	11 × 8 = 88	12 × 8 = 96
7 × 9 = 63	8 × 9 = 72	9 × 9 = 81	10 × 9 = 90	11 × 9 = 99	12 × 9 = 108
7 × 10 = 70	8 × 10 = 80	9 × 10 = 90	10 × 10 = 100	11 × 10 = 110	12 × 10 = 120
7 × 11 = 77	8 × 11 = 88	9 × 11 = 99	10 × 11 = 110	11 × 11 = 121	12 × 11 = 132
7 × 12 = 84	8 × 12 = 96	9 × 12 = 108	10 × 12 = 120	11 × 12 = 132	12 × 12 = 144

Start here

Days in the month

Thirty days have September, April, June and November.
All the rest have thirty-one, except February alone...
which has twenty-eight days clear, and twenty-nine each leap year

Answers - Homework Year 3

TERM 1

UNIT 1 NUMBER & ALGEBRA

Number Patterns and Sequences

1 10, 12, 14, 16. Up in 2s
2 15, 18, 21, 24. Up in 3s
3 9, 6, 3, 0. Down in 3s
4 42, 48, 54, 60. Up in 6s
5 100, 105, ... 120, 125. Up in 5s
6 500, 505, ... 515, 520. Up in 5s
7 301, 299, 297, 295. Down in 2s
8 980, 982, ... 986, 988. Up in 2s
9 99, 95, ... 87, 83. Down in 4s
10 43, 53, 63 The ones stay the same. The tens go up by 1 ten.
11 65, 85, 105 The ones stay the same. The tens go up by 2 tens.
12 26, 16, 6 The ones stay the same. The tens go down by 1 ten.
13 43, 32, 21 The ones go down by 1 one. The tens go down by 1 ten.
14 72, 62, 52 The ones stay the same. The tens go down by 1 ten.
15 103, 112, 121 The ones go down by 1 one. The tens go up by 1 ten.
16 Adult to check

Quick Check

1 100
2 110
3 191
4 991
5 1000
6 100
7 189
8 299
9 999
10 9999

Addition: Bridging through 10

1 23 using 3 + 3
2 36 using 1 + 6
3 34 using 2 + 4
4 45 using 3 + 5
5 56
6 83
7 64
8 76
9 91
10 55
11 83
12 45
13 53

UNIT 1 STATISTICS & PROBABILITY

Probability in the Bag

1 Red: unlikely
2 Yellow: impossible
3 Green: even chance
4 Blue: quite likely
5 8

UNIT 1 MEASUREMENT & SPACE

Position

1 J
2 E
3 K
4 F
5 B
6 H
7 K
8 N
9 P
10 C
11 Sample answer: Go 2 spaces to the right and then 2 spaces down.

UNIT 1 NAPLAN*-STYLE QUESTIONS

1 8 oranges
2 J
3 mint

UNIT 1 PROBLEM OF THE WEEK

Patterns with Numbers

Pattern 1: 3, 6, 9, 12, 15, 18, 21, 24
Pattern 2: 2, 5, 8, 11, 14, 17, 20, 23
Odd number out is 25.
Pattern 3: 31, 41, 51, 61, 71, 81, 91, 101
Pattern 4: 35, 40, 45, 50, 55, 60, 65, 70
Pattern 5: 18, 21, 24, 27, 30, 33, 36, 39
Odd number out is 43.

UNIT 2 NUMBER & ALGEBRA

Fractions

1 2, halves
2 3, thirds
3 4, quarters
4 6, sixths
5 8, eighths
6 one-half, $\frac{1}{2}$
7 two-thirds, $\frac{2}{3}$
8 three-quarters, $\frac{3}{4}$
9 two-sixths, $\frac{2}{6}$
10 three-eighths, $\frac{3}{8}$
11 $\frac{5}{4}$ or $1\frac{1}{4}$
12 $\frac{4}{3}$ or $1\frac{1}{3}$

Quick Check

1 one-quarter
2 five-eighths
3 two-thirds
4 two-fifths
5 three-quarters
6 five-sixths
7 $\frac{1}{3}$
8 $\frac{2}{5}$
9 $\frac{3}{4}$
10 $\frac{7}{8}$

Addition and Subtraction

1 13, 13 − 6 = 7, 13 − 7 = 6
2 17, 17 − 8 − 9, 17 − 9 = 8
3 24, 24 − 18 = 6, 24 − 6 = 18
4 21, 21 − 14 = 7, 21 − 7 = 14
5 25, 25 − 9 − 16, 25 − 16 = 9
6 45, 45 − 11 = 34, 45 − 34 = 11
7 82, 82 − 73 = 9, 82 − 9 = 73
8 93, 93 − 88 = 5, 93 − 5 = 88
9 102, 102 − 6 = 96, 102 − 96 = 6
10 130, 130 − 26 = 104, 130 − 104 = 26

UNIT 2 STATISTICS & PROBABILITY

Data Collection

1 Friday
2 Thursday
3 Mandarins
4 Pears
5 Tuesday and Wednesday
6 Monday and Tuesday
7 2 pieces of fruit
8 86

UNIT 2 MEASUREMENT & SPACE

Measurement Units

1 centimetres – ribbon – ruler
2 grams – apple – kitchen scales
3 litres – fish tank – bucket
4 millilitres – medicine bottle – measuring cup
5 kilograms – suitcase – bathroom scales
6–10 Adult to check estimates.
6 grams
7 centimetres
8 millilites
9 metres
10 litres

UNIT 2 NAPLAN*-STYLE QUESTIONS

1 3
2 Ginger
3 4 blocks

UNIT 2 PROBLEM OF THE WEEK

Fractions with Fish

1 13
2 Max
3 Luis
4 58
5 Open-ended
6 Open-ended

UNIT 3 NUMBER & ALGEBRA

Place Value

1 10
2 100
3 15
4 15
5 15
6 10
7 101
8 0, 10, 1
9 12, 0, 0
10 1, 6, 54
11 12, 3, 6
12 7, 0, 63
13 8, 6, 4
14 10, 63, 6
15 12, 0, 5

Quick Check

1 305
2 6160
3 170
4 5120
5 9300
6 1060

Bridging back through 10

1 23 − 3 − 5 = 15
2 44 − 4 − 2 = 38
3 42 − 2 − 4 = 36
4 83 − 3 − 5 = 75
5 43
6 17
7 26
8 55

ANSWERS

UNIT 3 STATISTICS & PROBABILITY

Max's Food Surveys

1 Three more people do not like broccoli than do like broccoli.

2

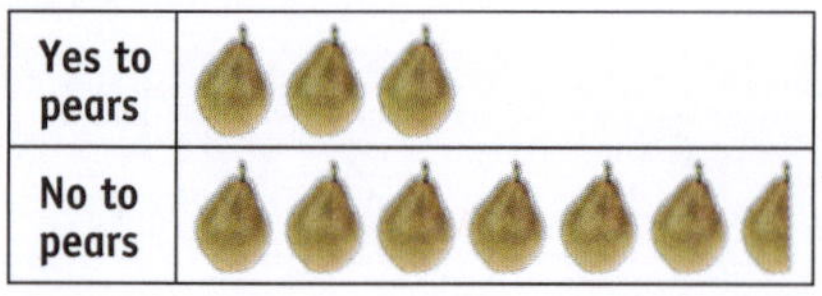

UNIT 3 MEASUREMENT & SPACE

3D Objects

3D objects	Faces	Edges	Vertices
Cube	6	12	8
Cylinder	3	2	0
Triangle-based pyramid	4	6	4
Square-based pyramid	5	8	5
Sphere	1	0	0

UNIT 3 NAPLAN*-STYLE QUESTIONS

1 150
2 True, False, False
3 A rectangular prism and a cube

UNIT 3 PROBLEM OF THE WEEK

Problem of the Week

1 21
2 4
3 6
4 Only 3 people played hockey.
5 Adult to check
6 Adult to check

UNIT 4 NUMBER & ALGEBRA

Place Value

1 One thousand, one hundred and twenty-one
1121
2 Two thousand, two hundred and one
2201
3 4000
4 4

5–10 Answers will vary. Sample answers:

5 1021, 1012 and 2011
6 2101, 1201 and 1102
7 2002, 2020 and 2200
8 211
9 1120
10 1003
11 Three thousand, four hundred and five
12 Four thousand, three hundred and sixteen
13 Fifteen thousand, five hundred and forty-six
14 Fourteen thousand and thirty-nine

Quick Check

1 5006   2 3017   3 15 670   4 4999   5 14 990

Make to 10

1 5 + 6 + 5 = 5 + 5 + 6 = 16
2 3 + 4 + 7 = 3 + 7 + 4 = 14
3 8 + 2 + 5 = 15
4 13 + 7 + 6 = 26
5 26 + 4 + 5 = 35

UNIT 4 STATISTICS & PROBABILITY

Bindi's Reading Chart

1 Thursday
2 Tuesday
3 15
4 28 + 20 + 5 + 40 + 15 = 108
5 Wednesday and Friday

UNIT 4 MEASUREMENT & SPACE

Measurement of Length

1 3 cm
2 2 cm
3 5 cm
4 10 cm
5 Adult to check
6–10 Adult to check

UNIT 4 NAPLAN*-STYLE QUESTIONS

1 0   2 5 days   3 $15

UNIT 4 PROBLEM OF THE WEEK

Birds at the Bird Bath

1 143
2 Pigeons
3 18
4 Adult to check

UNIT 5 NUMBER & ALGEBRA

Making Money Amounts

1 $1   3 $2   5 $1   7 $5·05   9 $2·30
2 $1   4 $1·15   6 $4·35   8 95c   10 $5·05

Quick Check

1 65c
2 35c
3 $2·30
4 $1·15
5 $1·05

6–10 More than one answer is possible. Sample answers:
6 20c, 20c, 20c
7 $1, 10c, 10c
8 $2, 5c, 5c
9 $1, 50c, 10c
10 $2, $1, 5c

Adding by Partitioning

1 27 + 10 + 2 = 39
2 33 + 30 + 5 = 68
3 46 + 40 + 3 = 89
4 29
5 41
6 26
7 48
8 77
9 58
10 28
11 107
12 101
13 109
14 110
15 119

UNIT 5 STATISTICS & PROBABILITY

Coloured Counters

	Impossible	Quite likely	Very likely
1	✓		
2			✓
3		✓	
4	✓		
5		✓	

UNIT 5 MEASUREMENT & SPACE

In the Art Room

1 2 m
2 3 m
3 2 m
4 3 m, left, $1\frac{1}{2}$ m, right, $2\frac{1}{2}$ m
5 Adult to check

UNIT 5 NAPLAN*-STYLE QUESTIONS

1 21   2 Gym   3 Likely

UNIT 5 PROBLEM OF THE WEEK

A New Reptile Park

Possible solution:

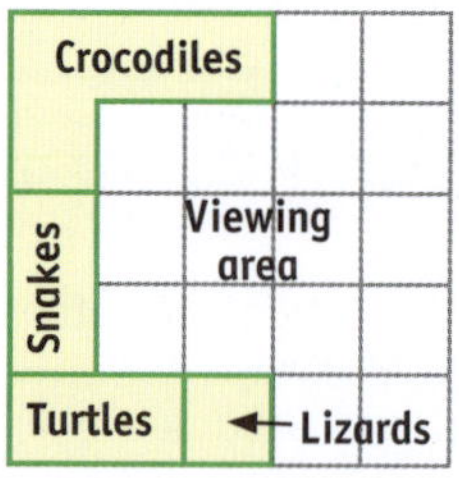

UNIT 6 NUMBER & ALGEBRA

Counting Sequences

1 12, 15, 18, 21, 24, 27, 30, 30, 3
2 16, 20, 24, 28, 32, 36, 40, 40, 4
3 20, 25, 30, 35, 40, 45, 50, 50, 5
4 40, 50, 60, 70, 80, 90, 100, 100, 10
5 The units number is always 0.
6 30 will make 3 groups of 10 or 10 groups of 3.
7 10
8 20
9 60
10 90

ANSWERS

TARGETING MATHS HOMEWORK: YEAR 3 © PASCAL PRESS ISBN: 9781925726565

Quick Check

1 5 2 5 3 5 4 5 5 5 6 20 7 18 8 35 9 70 10 24

Subtracting by Partitioning

1 27 − 10 − 5 = 12
2 58 − 30 − 4 = 24
3 46 − 20 − 3 = 23
4 53
5 17
6 31
7 54
8 73
9 71
10 74
11 61
12 63
13 45
14 35
15 22

UNIT 6 STATISTICS & PROBABILITY

Food Survey

Sample answers:

1 Very unlikely
2 Very likely
3 Unlikely
4 Impossible
5 Likely

UNIT 6 MEASUREMENT & SPACE

Angles

Adult to check

UNIT 6 NAPLAN*-STYLE QUESTIONS

1 50c, 50c, 20c, 5c
2 Fourth angle
3 Red

UNIT 6 PROBLEM OF THE WEEK

As Close to Zero as Possible

1 13

There are many different ways of reaching the closest possible number to zero. An example that uses the suggestions is to make the three numbers 54, 32 and 10 to get:

100 − 54 = 46
46 − 32 = 14
14 − 10 = 4

With the digits 0–6, make the three numbers 45, 36 and 20 and leave 1 to be added at the end to get:

100 − 45 = 55
55 − 36 = 19
19 − 20 = −1
−1 + 1 = 0

There are 36 possible ways of achieving this result. All of them involve having 4, 3 and 2 in the 10s positions and 6, 5 and 0 in the 1s position which leaves 1 to added as the last step.

UNIT 7 NUMBER & ALGEBRA

Multiplication

1 3 jumps of 5 = 15, 3 × 5 = 15
5 jumps of 3 = 15, 5 × 3 = 15
2 3 jumps of 4 = 12, 3 × 4 = 12
2 jumps of 6 = 12, 2 × 6 = 12
3 5, 10, 15, 20, 25, 30, 40
4 4, 8, 12, 16, 20, 40
5 6, 9, 12, 15, 18, 30
6 7, 49
7 10, 20, 30, 40
8 13

Quick Check

1 6 2 20 3 6 4 6 5 12 6 20 7 12 8 40 9 12 10 40 11 24 12 28 13 30 14 24

Multiplication Four Ways

1 3 + 3 + 3 = 9, 3 × 3 = 9
2 4 + 4 + 4 + 4 = 16, 4 × 4 = 16
3 5 + 5 + 5 + 5 + 5 + 5 = 30, 6 × 5 = 30
4 3 + 3 + 3 + 3 + 3 = 15, 5 × 3 = 15
5 4 × 3 = 12
6 3 × 2 = 6
7 4 × 10 = 40
8 5 × 5 = 25
9 5 × 2 = 10

UNIT 7 STATISTICS & PROBABILITY

Will It Happen?

Adult to check

UNIT 7 MEASUREMENT & SPACE

Making Models

1 12, 8 2 9, 6 3 8, 5 4 6, 4 5 12, 8

UNIT 7 NAPLAN*-STYLE QUESTIONS

1 20 2 12 3 White

UNIT 7 PROBLEM OF THE WEEK

Target 100: A Multiplication Game

Max's scores were: 4 × 2 = 8, 6 × 5 = 30, 5 × 4 = 20, 3 × 6 = 18.
This gives a total of 76, leaving 24 to go. Max needs a 4 and 6 to bring his score for five rounds to 100.
Kiku's scores were: 1 × 3 = 3, 4 × 6 = 24, 5 × 5 = 25, 6 × 6 = 36.
This gives a total of 88, leaving 12 to go. Kiku needs a 3 and 4 or a 2 and 6 to bring her score for five rounds to 100.
Kiku is the more likely winner as she has more ways to complete her score.

UNIT 8 NUMBER & ALGEBRA

Fractions of a Quantity

1 4 groups of 2, 2
2 3 groups of 3, 3
3 5 groups of 2, 2
4 2 groups of 6, 6
5 3 groups of 6, 6
6 3 of the 4 groups of 2, 6
7 3 groups of 4, 4
8 2 groups of 4, 4
9 5 groups of 2, 2
10 2 of the 3 groups of 3, 6

Quick Check

1 3 2 2 3 2 4 3 5 12 6 4 7 4 8 7 9 3 10 3

Multiplication Word Problems

1 15 blueberries
2 21 bananas
3 18 dollars
4 60 poppies
5 10 hours

Multiplication Families

Adult to check

UNIT 8 STATISTICS & PROBABILITY

School Travel Survey

Bus 6, Cycle 6, Car 10, Walk 12.
Adult to check chart.

UNIT 8 MEASUREMENT & SPACE

Telling the Time

1

2

3

4

5

6

7

8

9 60 minutes
10 24 hours
11 60 seconds
12 30 seconds
13 15 minutes

ANSWERS

TARGETING MATHS HOMEWORK: YEAR 3 © PASCAL PRESS ISBN: 9781925726565

UNIT 8 NAPLAN*-STYLE QUESTIONS

1 5 cherries 2 Third clock 3 2 birds

UNIT 8 PROBLEM OF THE WEEK

Counters in the Bag

1 Luis had 6 yellow counters.
2 Max had 5 green counters.
3 Kiku had 3 blue counters.
4 Luis had the most blue counters (8).
5 Adult to check
6 Adult to check

TERM 1 REVIEW

Review: Unit 1 and Unit 2

1 18, 20, 22, 24, 26, 28
2 19, 17, 15, 13, 11, 9
3 9, 12, 15, 18, 21, 24
4 101, 99, 97, 95, 93, 91
5 1000
6 1100
7 1410
8 16 + 4 + 3 = 23
9 28 + 2 + 4 = 34
10 Yellow
11 Red
12 Cross
13 Tick
14 Hexagon
15 Plus
16 $\frac{1}{4}$
17 $\frac{1}{3}$
18 $\frac{7}{8}$
19 13 − 4 = 9, 13 − 9 = 4
20 5 cm, 54 mm

Review: Unit 3 and Unit 4

1 11
2 13
3 3420
4 5116
5 1221
6 33 − 3 − 4 = 26
7 52 − 2 − 4 = 46
8 94 − 4 − 4 = 86
9 103 − 3 − 4 = 96
10 14
11 4th
12 2nd and 3rd
13 B
14 C
15 A

Review: Unit 5 and Unit 6

1 $1·80
2 $1·50
3 $4·05
4 55 − 20 − 3 = 32
5 72 − 30 − 8 = 34
6 up 4
7 right 3
8 down 2
9 left 1
10 down 2
11 12, 15, 18. 6 × 3 = 18
12 16, 20, 24. 6 × 4 = 24
13 9 times
14 9 times
15 Likely

Review: Unit 7 and Unit 8

1 15, 20
2 12, 20, 24
3 9, 12, 15, 18, 24
4 12
5 30
6 21
7 6, 8, 12
8 5, 5, 8
9 3, 0, 2
10 3 km
11 8 km
12 10 km
13 14 km
14 35 km
15 6:30, half past six
16 8:15, quarter past eight
17 3:45, quarter to four

TERM 2

UNIT 9 NUMBER & ALGEBRA

Estimation

1 Between 30 and 40
2 36
3 Between 40 and 50
4 48

Using a Calculator

5 18
6 57
7 154
8 6
9 27
10 52

Quick Check

1 29
2 105
3 110
4 12
5 136
6 64
7 94
8 37

Estimating Money Amounts

2 Between $8 and $9
3 Between $20 and $22
4 Between $14 and $15
5 Between $60 and $63
6 Yes
7 Yes
8 No
9 No
10 Yes

UNIT 9 STATISTICS & PROBABILITY

Red or Black

Outcomes

	Red	Black
1	0	6
2	1	5
3	2	4
4	3	3
5	4	2
6	5	1
7	6	0

8 Outcome 3 Red, 3 Black
9 Outcomes 6 Red, 0 Black and 0 Red, 6 Black
10 Only 1 way of achieving that outcome. All others have more than one possible way of happening.

UNIT 9 MEASUREMENT & SPACE

Turn and Turn Again

1 

2 

3 

4 

5 

6 

7 500 g
8 50 cm
9 250 g
10 25 cm

UNIT 9 NAPLAN*-STYLE QUESTIONS

1 250 g
2 60 pages
3 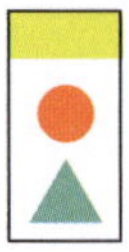

UNIT 9 PROBLEM OF THE WEEK

Choosing an End

Open-ended but good for discussion.

UNIT 10 NUMBER & ALGEBRA

Fractions

1 $\frac{3}{4}$ and $\frac{1}{4}$
2 $\frac{2}{4}$ and $\frac{2}{4}$
3 $\frac{3}{8}$ and $\frac{5}{8}$
4 $\frac{1}{3}$ and $\frac{2}{3}$
5 $\frac{3}{6}$ and $\frac{3}{6}$
6 $\frac{1}{5}$ and $\frac{4}{5}$
7 
8 
9 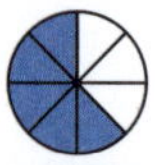
10 

ANSWERS

TARGETING MATHS HOMEWORK: YEAR 3 © PASCAL PRESS ISBN: 9781925726565

Quick Check

1 $\frac{5}{6}$ 3 $\frac{1}{3}$ 5 $\frac{1}{4}$ 7 $\frac{7}{10}$ 9 $\frac{1}{5}$
2 $\frac{1}{6}$ 4 $\frac{1}{8}$ 6 $\frac{2}{5}$ 8 $\frac{3}{5}$ 10 $\frac{4}{6}$

Money Amounts

1 $1, 50c, 50c
2 50c, 50c, 50c
3 50c, 20c
4 $1, $1, $1, 50c
5 50c, 20c, 20c, 5c
6 $2, $1, 50c
7 $2, $2, $2, $1, 50c, 10c
8 $1, $1, $1, 50c, 20c, 20c
9 $2, $2, 50c, 20c, 10c
10 $2, $2, 50c

UNIT 10 STATISTICS & PROBABILITY

Changing Chances

Sample answers:

	Impossible	Unlikely	Likely	Very Likely	Certain
1		✓			
2		✓			
3					✓
4				✓	
5	✓				
6		✓			

UNIT 10 MEASUREMENT & SPACE

Making a Model

1
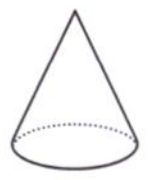
cone

2
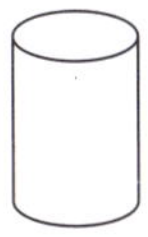
cylinder

3
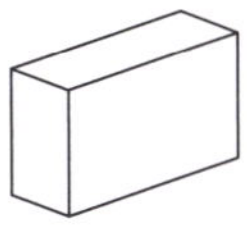
rectangular prism

4
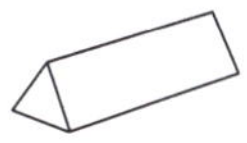
triangular prism

UNIT 10 NAPLAN*-STYLE QUESTIONS

1 5 2 $8 3 Sphere

UNIT 10 PROBLEM OF THE WEEK

Equal Shares – No Leftovers

Family Size	Packet of 11 Biscuits
2	$5\frac{1}{2}$ biscuits
3	$3\frac{2}{3}$ biscuits
4	$2\frac{3}{4}$ biscuits
5	$2\frac{1}{5}$ biscuits

Family Size	Packet of 13 Biscuits
2	$6\frac{1}{2}$ biscuits
3	$4\frac{1}{3}$ biscuits
4	$3\frac{1}{4}$ biscuits
5	$2\frac{3}{5}$ biscuits

The remaining part of this Problem of the Week is open-ended.

UNIT 11 NUMBER & ALGEBRA

Place Value

1 15 thousands and 34
2 60 thousands, 4 tens and 7 ones
3 65 thousands and 6 ones
4 11 110
5 56 hundreds, 7 tens and 4 ones
6 3656
7 3606
8 17 055
9 3606
10 1705

Quick Check

1 64 3 100 5 100 7 2350 9 8100
2 64 4 1000 6 146 8 660 10 12 120

Partitioning for Addition

1 300 + 90 + 8 = 398
2 300 + 90 + 9 = 399
3 600 + 80 + 9 = 689
4 200 + 90 + 7 = 297
5 300 + 80 + 7 = 387
6 900 + 90 + 9 = 999
7 800 + 40 + 9 = 849
8 300 + 0 + 8 = 308
9 300 + 30 + 9 = 339
10 1000 + 80 + 7 = 1087

UNIT 11 STATISTICS & PROBABILITY

School-night Bedtimes

1 53
2 8:30–9:00
3 9:30–10:00
4 31
5 14

UNIT 11 MEASUREMENT & SPACE

About Time

1 8:30

45 minutes

2 7:15

2 hours
15 minutes

3

3:30
4 hours
45 minutes

4 3

UNIT 11 NAPLAN*-STYLE QUESTIONS

1 20 324 2 Quarter past four 3 45

UNIT 11 PROBLEM OF THE WEEK

The 600-page Readathon

Recommendations for Week 3 and 4 should be:

	Last Week	This Week	Week 3	Week 4
Ada	226	134	120	120
Bindi	242	127	115	116
Luis	243	72	142	143
Max	140	140	160	160
Kiku	280	280	20	20

UNIT 12 NUMBER & ALGEBRA

On an Empty Number Line

Annotated number lines should resemble these lines.

1
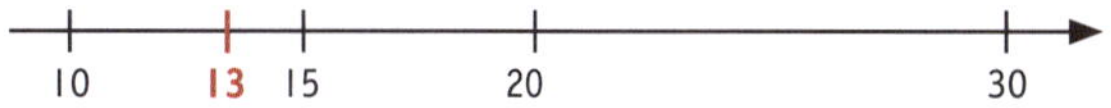

2
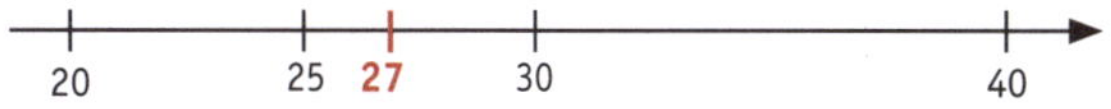

3
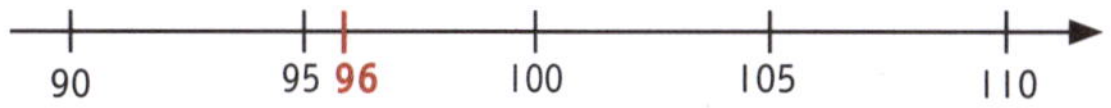

4
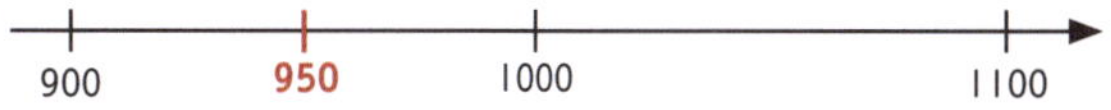

5
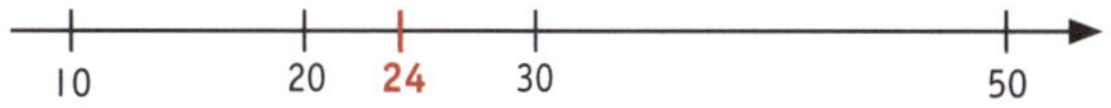

Quick Check

1 12 4 125 7 18 10 35 13 1250
2 33 5 175 8 700 11 100 14 10 000
3 19 6 12 9 77 12 141 15 10 500

Partitioning for Subtraction

1 100 + 50 + 2 = 152
2 100 + 20 + 1 = 121
3 50 + 2 = 52
4 400 + 10 + 1 = 411
5 100 + 10 + 1 = 111
6 200 + 20 + 6 = 226
7 100 + 20 + 3 = 123
8 200 + 20 + 4 = 224
9 100 + 10 = 110
10 500 + 20 + 3 = 523

UNIT 12 STATISTICS & PROBABILITY

Spinners

1 Open-ended, but Red will be close to 16, Blue to 6 and Yellow to 8.
2 Red
3 Blue
4 No
5 Because the calculated chances are only a guide to what might happen in the real world.

UNIT 12 MEASUREMENT & SPACE

Turtle Moves

1 Right 90°
Forward 2
Right 90°
Forward 2

2 Right 90°
Forward 4
Right 90°
Forward 4
Right 90°
Forward 4
Right 90°
Forward 4

UNIT 12 NAPLAN*-STYLE QUESTIONS

1 242
2 First bag
3 First option

UNIT 12 PROBLEM OF THE WEEK

Who Sat Where?

They were arranged clockwise around the table in this order: Luis, Hugo, Max, Kiku, Ada and Bindi.

UNIT 13 NUMBER & ALGEBRA

Sequencing and Ordering of Large Numbers

1 472 327
2 460 927
3 451 327
4 381 327
5 461 027
6 471 327

Quick Check

1 True
2 True
3 True
4 True
5 False
6 False
7 False
8 True
9 True
10 False

Extending Doubles Facts

1	8	80	800	8000	80 000
2	16	160	1600	16 000	160 000
3	32	320	3200	32 000	320 000
4	12	120	1200	12 000	120 000
5	24	240	2400	24 000	240 000
6	48	480	4800	48 000	480 000
7	30	300	3000	30 000	300 000
8	60	600	6000	60 000	600 000
9	120	1200	12 000	120 000	1 200 000

UNIT 13 STATISTICS & PROBABILITY

The Pizza Party

1 30
2 Meat Lovers
3 Vegetarian
4 Hawaiian
5 Cheese and Tomato and Supreme
6 Adult to check

UNIT 13 MEASUREMENT & SPACE

Measuring Tools

1 500 mL
2 1·5 kg
3 Up to 20 cm
4 Tape measure
5 125 mL
6 Adult to check
7 Scales
8 8
9 Between 15 cm and 20 cm
10 Adult to check

UNIT 13 NAPLAN*-STYLE QUESTIONS

1 42
2 Labels are 0, 4, 8 and 12
3 7 cm

UNIT 13 PROBLEM OF THE WEEK

Cordial Recipe

1 300 mL
2 See diagram
3 Adult to check
4 Adult to check

UNIT 14 NUMBER & ALGEBRA

Multiplication with Tallies

1 4 × 5 + 4 × 3 = 20 + 12 = 32
2 6 × 5 + 6 × 2 = 30 + 12 = 42
3 4 × 5 + 4 × 4 = 20 + 16 = 36
4 6 × 5 + 6 × 3 = 30 + 18 = 48
5 5 × 5 + 5 × 3 = 25 + 15 = 40
6 4 × 5 + 4 × 2 = 20 + 8 = 28
7 8 × 5 + 8 × 1 = 40 + 8 = 48
8 5 × 5 + 5 × 4 = 25 + 20 = 45

Quick Check

1 21
2 24
3 27
4 35
5 42
6 24
7 45
8 48
9 18
10 32
11 33
12 44
13 55
14 66
15 77

Multiplication Word Problems

1 6 × 8 = 48
2 5 × 6 = 30
3 20c × 8 = $1·60
4 5 × 9 = 45
5 6 × 7 = 42

UNIT 14 STATISTICS & PROBABILITY

Favourite Movies

Sample answers:
- Do you like cartoon movies about animals?
- Do you like movies about people?
- Do you like adventure movies?

UNIT 14 MEASUREMENT & SPACE

Paper Folding to Make Angles

1 16
2 4
3 6
4 6
5 9 (including the right angles)

UNIT 14 NAPLAN*-STYLE QUESTIONS

1 $25
2 Less than a right angle
3 42

UNIT 14 PROBLEM OF THE WEEK

Number Bingo

Two of the blank number clues need to be 7 × 5 to cover 35 and 2 × 9 or 6 × 3 to cover 18. The third blank number clue is open-ended so long as it does not make a match with one of the other totals. Also, the bingo cards need 21 and 28 on two different cards to match the number clues 4 × 2 and 3 × 7.

UNIT 15 NUMBER & ALGEBRA

Multiplication and Division

1 4 × 6 = 24, 6 × 4 = 24, 24 ÷ 6 = 4, 24 ÷ 4 = 6
2 5 × 7 = 35, 7 × 5 = 35, 35 ÷ 7 = 5, 35 ÷ 5 = 7
3 3 × 9 = 27, 9 × 3 = 27, 27 ÷ 9 = 3, 27 ÷ 3 = 9
4 6 × 7 = 42, 7 × 6 = 42, 42 ÷ 7 = 6, 42 ÷ 6 = 7
5 7 × 3 = 21, 3 × 7 = 21, 21 ÷ 3 = 7, 21 ÷ 7 = 3
6 8 × 3 = 24, 3 × 8 = 24, 24 ÷ 3 = 8, 24 ÷ 8 = 3
7 4 × 9 = 36, 9 × 4 = 36, 36 ÷ 9 = 4, 36 ÷ 4 = 9
8 7 × 8 = 56, 8 × 7 = 56, 56 ÷ 8 = 7, 56 ÷ 7 = 8
9 9 × 7 = 63, 7 × 9 = 63, 63 ÷ 7 = 9, 63 ÷ 9 = 7

TARGETING MATHS HOMEWORK: YEAR 3 © PASCAL PRESS ISBN: 9781925726565

Quick Check

1 3 × 8 = 24
2 8 × 2 = 16
3 5 × 4 = 20
4 3 × 6 = 18
5 3 × 7 = 21
6 10 × 9 = 90
7 3 × 9 = 27
8 4 × 4 = 16
9 5 × 5 = 25
10 9 × 4 = 36

Division Word Problems

1 12 ÷ 3 = 4
2 20 ÷ 5 = 4
3 16 ÷ 4 = 4
4 15 ÷ 3 = 5
5 21 × 2 = 42, 42 ÷ 3 = 14

UNIT 15 STATISTICS & PROBABILITY

Bean Growth

1 Luis's bean was 10 cm tall.
2 Max's bean was 15 cm tall.
3 15 cm, 15 cm
4 25 cm, 25 cm
5 Open-ended

UNIT 15 MEASUREMENT & SPACE

What Is the Time?

1 24 hours
2 12 hours
3 12 hours
4 A
5 C
6 B
7 5 o'clock or 5 am
8 3 o'clock or 3 pm
9 half past 7 or 7:30 pm
10 Adult to check

UNIT 15 NAPLAN*-STYLE QUESTIONS

1 4
2 12
3 45 cm

UNIT 15 PROBLEM OF THE WEEK

Party Balloons

1 Yes, 24
2 Yes, 18
3 No
4 The factors of 72: 1, 2, 3, 4, 6, 8, 9, 12, 18, 24, 36, 72

UNIT 16 NUMBER & ALGEBRA

Working Smart Not Hard for Addition and Subtraction

1 48, 58, 37
2 51, 61, 38
3 58, 68, 29
4 108, 118, 96
5 62, 72, 37
6 99, 109, 46
7 110, 120, 35
8 90, 100, 27
9 79, 89, 32
10 79, 99, 43

Quick Check

Answers:

1 67 + 29 = 96
2 38 + 27 = 65
3 46 + 24 = 70
4 96 + 12 = 108
5 103 + 26 = 129
6 46 + 34 = 80
7 96 + 22 = 118
8 93 + 26 = 119
9 77 + 29 = 106
10 48 + 37 = 85
11 139 − 26 = 113
12 70 − 46 = 24
13 96 − 29 = 67
14 118 − 22 = 96
15 65 − 27 = 38

Matches:

1, 9, 13
2, 10, 15
3, 6, 12
4, 7, 14
5, 8, 11

Estimation

1 Open-ended
2 Open-ended
3 Open-ended
4 23
5 16
6 22
7 Open-ended
8 Open-ended

UNIT 16 STATISTICS & PROBABILITY

Rolling a Cube

1 15
2 Green
3 3
4 2
5 1
6 Green
7 There are more green dots than blue or orange.

UNIT 16 MEASUREMENT & SPACE

Measuring Mass and Capacity

1 700 g
2 400 g
3 100 g
4 300 mL
5 500 mL
6 800 mL
7 350 g
8 300 g
9 500 mL
10 250 mL

UNIT 16 NAPLAN*-STYLE QUESTIONS

1 26 + 7 = ☐
2 100 g
3 36 fish

UNIT 16 PROBLEM OF THE WEEK

Raindrops

Apart from Q3 where the answer is 72, these questions are open-ended.

TERM 2 REVIEW

Review: Unit 9 and Unit 10

1 Open-ended
2 41
3–10 Sample answers provided
3 50
4 110
5 70
6 $13
7 $30
8 $12
9 $18
10 $7
11 $\frac{3}{4}$
12 $\frac{1}{6}$
13 $\frac{3}{8}$
14 $\frac{3}{10}$
15 $\frac{2}{3}$
16 $\frac{2}{5}$
17 $\frac{1}{2}$
18 $\frac{7}{10}$
19 $2, $1, 10c
20 Sample answer: 50c, 50c, 50c, 20c, 5c
21 Open-ended
22 Open-ended

Review: Unit 11 and Unit 12

1 1515
2 11 500
3 202
4 299
5 2579
6 13
7 Sample answer: Favourite Fruit
8 5:30

9 5:45

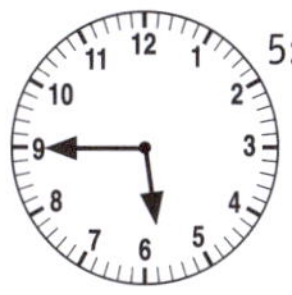

10 15 minutes
11 124
12 216
13 Red 11, Yellow 6, Blue 3

Review: Unit 13 and Unit 14

1 False
2 True
3 True
4 True
5 160
6 500 mL
7 15 cm
8 250 g
9 10 cm
10 36
11 21
12 55
13 24
14 6 × 7 = 42
15 30c × 5 = $1·50
16 3

Review: Unit 15 and Unit 16

1 4 × 3 = 12
2 3 × 5 = 15
3 4 × 6 = 24
4 27 ÷ 3 = 9
5 24 ÷ 3 = 8
6 17 cm
7 6 cm
8 12 hours
9 37 + 13 = 50, 37 + 23 = 60, 60 − 23 = 37
10 46 + 15 = 61, 56 + 15 = 71, 61 − 15 = 46
11 100 g
12 150 g

TERM 3

UNIT 17 NUMBER & ALGEBRA

Pattern Rules

1 8, 6, 9, 7
2 11, 17, 23, 29, 35, 41
3 8, 6, 12, 10, 20, 18
4 6, 12, 15, 30, 33, 66
5 5, 4, 8, 7, 11, 10
6 32, 64, 128. Double.
7 13, 14. Add 3, add 1.
8 4, 6, 5. Add 2, subtract 1.

ANSWERS

Quick Check

1 14, 15, 18
2 46, 57, 68
3 15, 11, 10
4 10, 9, 12
5 27, 33, 40
6 18, 26, 36
7 21, 26, 31
8 102, 106, 107, 110
9 100, 99, 102
10 102, 105, 108

Unknown Values

1 69 2 56 3 121 4 63 5 8 6 54 7 22 8 9 9 7 10 74 11 48 12 64

UNIT 17 STATISTICS & PROBABILITY

Making a Graph

	Coin	Number	Value
1	$2	9	$18
2	$1	6	$6
3	50c	4	$2
4	20c	3	$0·60
5	10c	9	$0·90
6	5c	12	$0·60
7	**Total**	**43**	**$28·10**

8 Max has 43 coins with a total value of $28·10.

UNIT 17 MEASUREMENT & SPACE

Estimating and Measuring Length

1 B 2 C 3 B, C 4 A, D 5 10 cm 6 5 cm 7 C, B 8 8 cm

UNIT 17 NAPLAN*-STYLE QUESTIONS

1 30 2 16 3 12 cm

UNIT 17 PROBLEM OF THE WEEK

How Long Are These Lines?

All the lines are 30 cm.

UNIT 18 NUMBER & ALGEBRA

Fractions

1 halves 2 thirds 3 quarters 4 eighths 5 $\frac{1}{2}$ 6 $\frac{1}{3}$ 7 $\frac{1}{4}$ 8 3

Quick Check

1 $\frac{3}{8}$ and $\frac{5}{8}$
2 $\frac{2}{6}$ and $\frac{4}{6}$
3 $\frac{2}{3}$ and $\frac{1}{3}$
4 $\frac{2}{5}$ and $\frac{3}{5}$
5 $\frac{2}{4}$ and $\frac{2}{4}$
6 $\frac{2}{6}$ and $\frac{4}{6}$
7 $\frac{1}{3}$ and $\frac{2}{3}$
8 $\frac{3}{8}$ and $\frac{5}{8}$
9 $\frac{2}{5}$ and $\frac{3}{5}$
10 No

Unknown Values in Addition

1 5 2 8 3 10 4 26 5 17 6 22 7 3 8 100 9 32 10 30 11 6 12 9 13 10 14 11 15 12 16 20 17 11 18 20 19 26 20 17

UNIT 18 STATISTICS & PROBABILITY

Data Collection

1 39 2 Swimming 3 15 4 113 5 35

UNIT 18 MEASUREMENT & SPACE

Money Amounts

1 20c + 10c
2 20c + 5c + 5c
3 10c + 10c + 5c + 5c
4 10c + 5c + 5c + 5c + 5c
5 5c + 5c + 5c + 5c + 5c + 5c
6 10c + 10c + 10c

UNIT 18 NAPLAN*-STYLE QUESTIONS

1 19 2 Second chart 3 15c

UNIT 18 PROBLEM OF THE WEEK

Who Had Which Coins?

Bindi has five 50c coins and five 10c coins.
Luis has two $1 coins and five 20c coins.
Max has one $2 coin, one 50c coin and five 10c coins.
or
Bindi has four 50c coins, four 20c coins and two 10c coins.
Luis has two $1 coins, one 50c coin, one 20c coin and three 10c coins.
Max has one $2 coin, one 50c coin and five 10c coins.

UNIT 19 NUMBER & ALGEBRA

Multiplication

1 9
2 12
3 15
4 18
5 +3 each time
6 69
7 72
8 75
9 78
10 12
11 16
12 20
13 24
14 132
15 136
16 140
17 144
18 15
19 25
20 35
21 45
22 115
23 125
24 135
25 145
26 30
27 60
28 90
29 120
30 130
31 160
32 190
33 220

Quick Check

1 30 2 30 3 18 4 18 5 30 6 30 7 24 8 24 9 70 10 44 11 35 12 36 13 42 14 21 15 130 16 140 17 28 18 20 19 33 20 48

Multiplication and Division

1 5 × 2 = 10, 10 ÷ 5 = 2
2 2 × 5 = 10, 10 ÷ 2 = 5
3 6 × 3 = 18, 18 ÷ 6 = 3
4 3 × 6 = 18, 18 ÷ 3 = 6

UNIT 19 MEASUREMENT & SPACE

Calendar Months

1 Wednesday
2 5th June
3 4th July
4 May
5 Winter (in the Southern Hemisphere)
6 April, September, November

UNIT 19 STATISTICS & PROBABILITY

Any Chance?

1 Adult to check
2 Unlikely
3 Unlikely
4 Likely
5 Impossible
6 Certain
7 Adult to check

UNIT 19 NAPLAN*-STYLE QUESTIONS

1 6 × 6
2 3 × 5 = 15, 15 ÷ 3 = 5
3 Wednesday

UNIT 19 PROBLEM OF THE WEEK

The Chef Challenge

1 3
2 Pizza should show 12 olives, 4 mozzarella balls, 6 salami slices and 5 pineapple chunks.
3 4
4 Pizza should show 6 mini-meatballs, 3 anchovy halves, 4 tomato slices and 8 olives.

UNIT 20 NUMBER & ALGEBRA

Counting Play Money

1 $411 2 $121 3 $300 4 $526 5 Ada 6 Kiku 7 $405 8 $111 9 $290 10 $1358

Quick Check

1 $416 2 $1165 3 $1025 4 $1332 5 $3001 6 $2365 7 $10 656 8 $1035 9 $21 760 10 $4906

TARGETING MATHS HOMEWORK: YEAR 3 © PASCAL PRESS ISBN: 9781925726565

Making Friendly Numbers for Addition

1 30 + 50 = 80
2 30 + 40 = 70
3 50 + 80 = 130
4 40 + 44 = 84
5 30 + 63 = 93
6 80 + 22 = 102
7 40 + 81 = 121
8 22 + 90 = 112
9 81 + 30 = 111
10 100 + 85 = 185

UNIT 20 STATISTICS & PROBABILITY

Counters in a Bag

1 2, 3, 4, 5, 6
2 Yes.
3 4

UNIT 20 MEASUREMENT & SPACE

Finding Shapes in the Environment

1–10 are open-ended questions.

UNIT 20 NAPLAN*-STYLE QUESTIONS

1 + 1 0 0 =
2 8
3 Rectangular prism

UNIT 20 PROBLEM OF THE WEEK

Closest to $5000

1 **At the end of three rounds:**
- Bindi had $2110.
- Luis had $2555.
- Max had $3130.
- Kiku had $3510.

2 Kiku
3 **Round 4**

Player	Current Score	Notes Needed	Final Score
Bindi	$2110	three $1000	$5110
Luis	$2555	two $1000, one $100	$4655
Max	$3130	two $1000, one $5	$5135
Kiku	$3510	one $1000, two $100	$4710

4 Bindi will be closest to $5000.
5 Luis will be furthest from $5000.

UNIT 21 NUMBER & ALGEBRA

Estimation

1 About 30
2 About 50
The number closest to a ten is rounded to that ten.
3 30 + 33 = 63
4 36 + 30 = 66
5 33 + 70 = 103
6 63 + 40 = 103
7 60 + 24 = 84
8 50 + 24 = 74
9 76 + 30 = 106
10 100 + 13 = 113

Quick Check

1 38
2 77
3 94
4 54
5 96
6 103
7 104
8 73
9 57
10 117

Addition Using Pairs That Make 10

1 9 + 1 = 10, 90 + 10 = 100, 900 + 100 = 1000
2 2 + 8 = 10, 20 + 80 = 100, 200 + 800 = 1000
3 7 + 3 = 10, 70 + 30 = 100, 700 + 300 = 1000
4 6 + 4 = 10, 60 + 40 = 100, 600 + 400 = 1000
5 5 + 5 = 10, 50 + 50 = 100, 500 + 500 = 1000
6 6 + 6 = 12, 60 + 60 = 120, 600 + 600 = 1200
7 7 + 7 = 14, 70 + 70 = 140, 700 + 700 = 1400
8 8 + 8 = 16, 80 + 80 = 160, 800 + 800 = 1600
9 9 + 9 = 18, 90 + 90 = 180, 900 + 900 = 1800
10 11 + 11 = 22, 110 + 110 = 220, 1100 + 1100 = 2200

UNIT 21 MEASUREMENT & SPACE

How Many Minutes?

1 25
2 24
3 59
4 6
5 12

UNIT 21 STATISTICS & PROBABILITY

Comparing Results

1 The Most Common Pets at School
2 Dogs
3 Bindi's
4 Luis's
5 Bindi 4, Luis 5
6 2
7 3

UNIT 21 NAPLAN*-STYLE QUESTIONS

1 40
2 7:20
3 28

UNIT 21 PROBLEM OF THE WEEK

Making Totals

There are many possible answers to these questions apart from making 14 and 15, where 14 = 6 + 5 + 3 and 15 = 6 + 5 + 4 are the only possibilities.

Including 7 and 8: Possible totals now include 20 = 8 + 7 + 5 and 21 = 8 + 7 + 6.

Including 9 and 10: Possible totals now include 26 = 10 + 9 + 7 and 27 = 10 + 9 + 8.

UNIT 22 NUMBER & ALGEBRA

Unknowns in Subtraction

1 22
2 25
3 36
4 36
5 35
6 45
7 62
8 60
9 31
10 6
11 6
12 8
13 12
14 10
15 2

Quick Check

1 answer, 34
2 change, 9
3 start, 60
4 answer, 99
5 start, 70
6 start, 48
7 answer, 89
8 change, 9
9 change, 14
10 change, 4

Patterns in Multiplication Tables

1 25, 30, 35, 40, 45, 50, 55, 60, 65, 70, 75, 80, 85, 90, 95, 100, 105, 110
2 No
3 Not ending in 0 or 5
4 Yes
5 Ends in 5
6 20, 24, 28, 32, 36, 40, 44, 48, 52, 56, 60, 64, 68, 72, 76, 80, 84, 88
7 No
8 Ends in an odd number
9 Yes
10 Ends in 24 which is in the counting sequence

UNIT 22 STATISTICS & PROBABILITY

Shapes in a Box

1 4
2 Shapes in the Shape Box
3

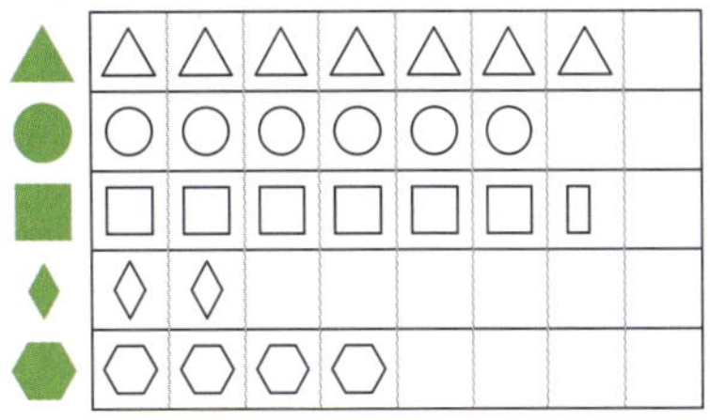

UNIT 22 MEASUREMENT & SPACE

Estimating Mass

Open-ended

UNIT 22 NAPLAN*-STYLE QUESTIONS

1 $5
2 3
3 3 apples

ANSWERS

UNIT 22 PROBLEM OF THE WEEK

Going Camping

There are 12 kg of items in total so each backpack should contain 3 kg. Here is a sample answer:

1 milk, cereal, sausages
2 milk, bread
3 orange juice, tomato sauce, burgers, bread
4 orange juice, apples, tomatoes, honey

UNIT 23 NUMBER & ALGEBRA

Multiplication

1 80
2 5 × 10 = 150
3 6 × 4 × 10 = 240
4 7 × 3 × 10 = 210
5 30 + 2, 180 + 12 = 192
6 20 + 7, 60 + 21 = 81
7 40 + 1, 120 + 3 = 123
8 30 + 8, 120 + 32 = 152
9 50 + 3, 250 + 15 = 265
10 40 + 2, 240 + 12 = 252

Quick Check

1 120
2 250
3 210
4 320
5 440
6 93
7 168
8 255
9 282
10 920

Multiplication Word Problems

1 24
2 3 × 8 = $24
3 6 × 4 = 24
4 5 × 3 = 15
5 24 × 5 = 120
6 10 × 7 = 70
7 3 × 22 = 66

UNIT 23 STATISTICS & PROBABILITY

Making a Column Graph

1 Insects in the School Garden
2 See chart
3

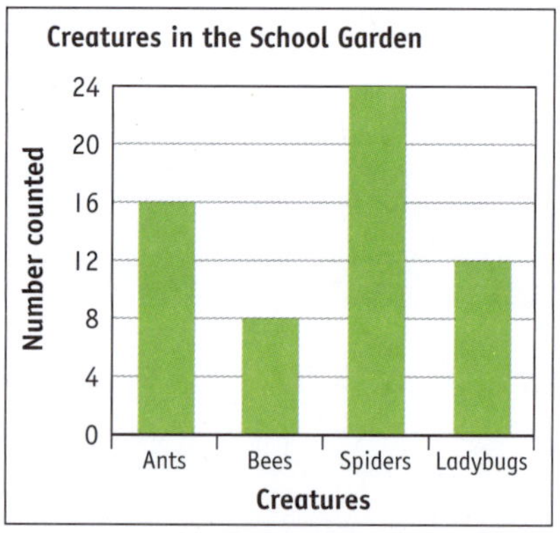

UNIT 23 MEASUREMENT & SPACE

A Lost Dog

1 Go north and turn right.
2 Go east to Park Road and turn right.
3 Turn left into School Road.
4 Turn right into Main Road.
5 Go east until you find the dog.

UNIT 23 NAPLAN*-STYLE QUESTIONS

1 32
2 2
3 30 km

UNIT 23 PROBLEM OF THE WEEK

Going Bush

8 packets of 4 bread rolls
4 packets of 16 cheese slices
4 packets of 8 tomatoes
96 bottles of iced water
16 apples
192 strawberries
64 mandarins
Adult to check class plan

UNIT 24 NUMBER & ALGEBRA

Fractions and Division

1 20, 25, 30, 35, 40, 45, 50
2 20 + 2, 10 + 1 = 11
3 20 + 8, 10 + 4 = 14
4 30 + 6, 15 + 3 = 18
5 20, 25, 30, 35, 40, 45, 50
6 30 + 6, 10 + 2 = 12
7 60 + 3, 20 + 1 = 21
8 90 + 6, 30 + 2 = 32

Quick Check

1 8
2 12
3 23
4 16
5 27
6 10
7 22
8 13
9 16
10 24

Division Word Problems

1 12 cherries
2 9 oranges
3 48 pages
4 $37
5 32 pages
6 8 strawberries
7 $3

UNIT 24 STATISTICS & PROBABILITY

Probability

1 Red
2 Purple and yellow
3 Blue
4 No
5–6 Yes because red is the largest.

UNIT 24 MEASUREMENT & SPACE

Half Full

1 1 L
2 500 mL
3 250 mL
4 325 mL
5 500 g
6 250 g
7 180 g
8 750 g

UNIT 24 NAPLAN*-STYLE QUESTIONS

1 13
2 8
3 $\frac{1}{2}$ kg box

UNIT 24 PROBLEM OF THE WEEK

Guess the Number Puzzles (With Division)

1 18
2 50
3 24
4 20
5 6
6–7 Open-ended

TERM 3 REVIEW

Review: Unit 17 and Unit 18

1 15, 19, 23, 27, 31, 35
2 21, 24, 27, 30, 33, 36
3 4, 6, 5, 7, 6
4 45, 56, 67, 78, 89, 100
5 100, 99, 104, 103, 108, 107
6 27
7 38
8 101
9 31
10 42
11 71
12 $19·50
13 5 cm
14 $\frac{1}{4}$
15 $\frac{1}{8}$
16 $\frac{1}{3}$
17 $\frac{1}{2}$
18 43
19 12
20 7
The last three answers can be in any order.
21 Group 1: 50c, 20c, 5c, 5c
22 Group 2: 50c, 10c, 10c, 10c
23 Group 3: 20c, 20c, 20c, 20c

Review: Unit 19 and Unit 20

1 12
2 18
3 35
4 32
5 18
6 35
7 72
8 55
9 5 × 3 = 15
10 3 × 5 = 15
11 Wednesday
12 3rd June
13 25th June
14 July
15 $461
16 $1221
17 $2615
18 $4010
19 $20 606
20 30 + 51 = 81
21 40 + 41 = 81
22 80 + 14 = 94
23 70 + 71 = 141
24 100 + 21 = 121
25 80 + 30 = 110

Review: Unit 21 and Unit 22

1 40 + 33 = 73
2 30 + 34 = 64
3 23 + 60 = 83
4 70 + 34 = 104
5 100 + 23 = 123
6 90 + 76 = 166
7 55 minutes
8 6 minutes
9 71 minutes
10 Tuesday
11 Tuesday
12 12
13 Daffodils
14 Gladioli
15 46
16 37
17 38
18 111
19 17
20 12
21 Adult to check

TARGETING MATHS HOMEWORK: YEAR 3 © PASCAL PRESS ISBN: 9781925726565

Review: Unit 23 and Unit 24

1 60
2 3 × 3 × 10 = 90
3 3 × 6 × 10 = 180
4 4 × 4 × 10 = 160
5 30 + 3, 180 + 18 = 198
6 20 + 5, 60 + 15 = 75
7 10 + 5, 70 + 35 = 105
8 5 per row
9 10 ladybugs
10 10 more bees
11 55
12 6 × 6 = 36 stickers
13 24 ÷ 4 = 6 stickers
14 80 + 6, 40 + 3 = 43
15 60 + 3, 20 + 1 = 21
16 40 + 8, 10 + 2 = 12
17 1 L jug has 400 mL

TERM 4

UNIT 25 NUMBER & ALGEBRA

Large Numbers

1 26, 327. Twenty-six thousand, three hundred and twenty-seven
2 38, 546. Thirty-eight thousand, five hundred and forty-six
3 27, 689. Twenty-seven thousand, six hundred and eighty-nine
4 39 thousands
5 43 thousands
6 29 thousands
7 60 thousands

Quick Check

1 26 472
2 38 208
3 40 650
4 55 050
5 100 001

Number Patterns

1 3, 6, 5, 8, 7, 10, 9, 12
The 10th number is 14.
2 2, 4, 3, 6, 5, 10, 9, 18
The 10th number is 34.
3 51, 45, 47, 41, 43, 37, 39, 33
The 10th number is 29
4 2, 6, 3, 9, 6, 18, 15, 45
The 10th number is 126.
5 90, 95, 94, 99, 98, 103, 102, 107
The 10th number is 111.
6 110, 105, 108, 103, 106, 101, 104, 99
The 10th number is 97.
7 Add 2, subtract 4.
8 Multiply by 2, subtract 2.
9 Add one more each time.
10 Subtract 2, add 4.

UNIT 25 STATISTICS & PROBABILITY

What's the Weather?

Open-ended, but suggested answers below.

1 Cloud, 21 °C
2 Rain, 20 °C
3 Quite likely
4 Unlikely
5 Very unlikely

UNIT 25 MEASUREMENT & SPACE

Time Lines

1 40 min + 10 min = 50 min
2 45 min + 5 min = 50 min
3 5 min + 15 min = 20 min
4 30 min + 15 min = 45 min
5 10 min + 30 min = 40 min

UNIT 25 NAPLAN*-STYLE QUESTIONS

1 25
2 97 650
3 6:30

UNIT 25 PROBLEM OF THE WEEK

Screen Time

These are suggestions only. There are many possible answers.

Max's Plan

Mon.	Tue.	Wed.	Thu.	Fri.	Sat.	Sun.	Total
40 min			40 min	40 min	1 h	1 h	4 h

Kiku's Plan

Mon.	Tue.	Wed.	Thu.	Fri.	Sat.	Sun.	Total
	1 h	1 h	1 h	20 min	20 min	20 min	4 h

Hugo's Plan

Mon.	Tue.	Wed.	Thu.	Fri.	Sat.	Sun.	Total
40 min	40 min	40 min		40 min	40 min	40 min	4 h

UNIT 26 NUMBER & ALGEBRA

Fractions

1 $\frac{3}{8}$
2 $\frac{5}{8}$
3 $\frac{2}{6}$ or $\frac{1}{3}$
4 $\frac{4}{6}$ or $\frac{2}{3}$
5 $\frac{4}{8}$ or $\frac{1}{2}$
6 $\frac{4}{8}$ or $\frac{1}{2}$
7 $\frac{4}{10}$ or $\frac{2}{5}$
8 $\frac{6}{10}$ or $\frac{3}{5}$
9 $\frac{1}{6}$
10 $\frac{5}{6}$
11 6
12 8
13 18

Quick Check

1 $\frac{1}{2}$
2 $\frac{3}{4}$
3 $\frac{7}{8}$
4 $\frac{2}{3}$
5 $\frac{4}{5}$
6 $\frac{1}{4}$
7 $\frac{5}{8}$
8 $\frac{5}{6}$
9 $\frac{3}{5}$
10 $\frac{1}{8}$

Bar Model for Fractions

1 Luis has 18 pencils. Max has 6 pencils.
2 Kiku has 7 stickers. Hugo has 15 stickers
3 Bindi has 18 pencils. Ada has 12 pencils
4 Max has 16 crayons. Kiku has 10 crayons
5 Hugo has 47 marbles. Ada has 17 marbles.

UNIT 26 STATISTICS & PROBABILITY

Make a Chart

UNIT 26 MEASUREMENT & SPACE

Perimeter

1 About 100 cm (or 1 m)
2 101.4 cm

UNIT 26 NAPLAN*-STYLE QUESTIONS

1 $\frac{3}{5}$
2 15
3 The same number of students hate kale or peas as hate sprouts.

UNIT 26 PROBLEM OF THE WEEK

The Giant Chocolate Bar

1 \$3
2 $\frac{1}{2}$
3 $\frac{1}{3}$
4 $\frac{1}{12}$
5 Luis had 24 pieces.
Max had 16 pieces.
Kiku had 4 pieces.
Hugo had 4 pieces.

ANSWERS

UNIT 27 NUMBER & ALGEBRA

Estimation

1 Estimate should be within 5 of 31.
2 Estimate should be within 5 of 27.
3 Estimate should be within 5 of 42.
4 Estimate should be within 5 of 36.

Quick Check

1 $700	5 $900	9 $4000	13 $60 000
2 $400	6 $7000	10 $3000	14 $40 000
3 $400	7 $2000	11 $10 000	15 $10 000
4 $600	8 $5000	12 $40 000	

Number Sentences for Word Problems

1 $53 + $36 = $89
2 30 × 7 = 210 minutes
3 15 ÷ 3 = 5 apples
4 36 − 9 = 27 rolls
5 6 × 22 = 132 cm

UNIT 27 STATISTICS & PROBABILITY

Dog-walking Income

1

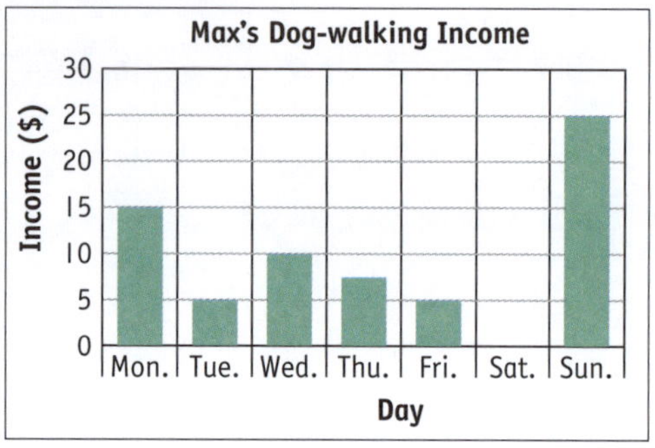

2 $5
3 Monday and Wednesday
4 Monday, Tuesday and Friday
5 $67·50

UNIT 27 MEASUREMENT & SPACE

3D Objects and Nets

1 4 2 1 3 6 4 3 5 5 6 2

UNIT 27 NAPLAN*-STYLE QUESTIONS

1 2000 2 40 3 10

UNIT 27 PROBLEM OF THE WEEK

The School Litter Problem

The answers are in the table below.

Rubbish items	Team 1	Team 2	Team 3	Total
Food wrappers	16		10	26
Recyclables	12	7	8	27
Food scraps	16	3	12	31
Lost property	2	3	4	9
Scrap paper	14	11	6	31
Other		4		4
				128

UNIT 28 NUMBER & ALGEBRA

Fraction Number Lines

1

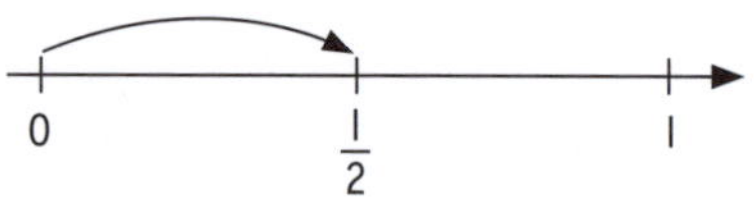

2

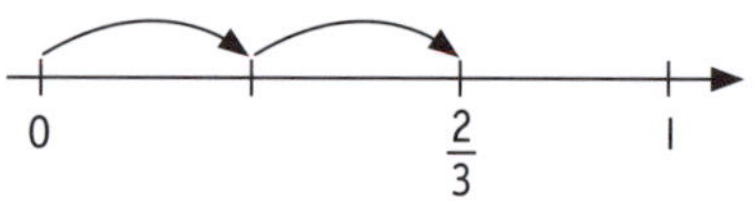

3

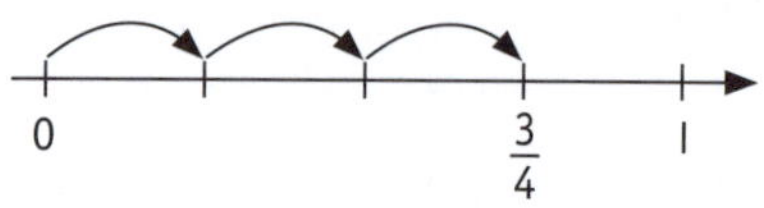

4

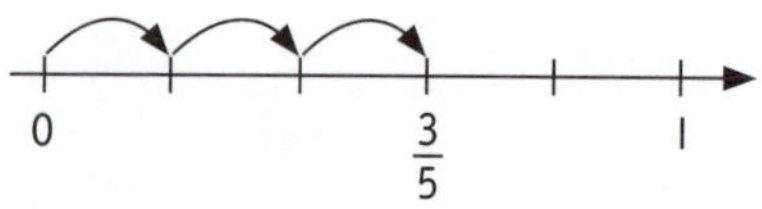

5

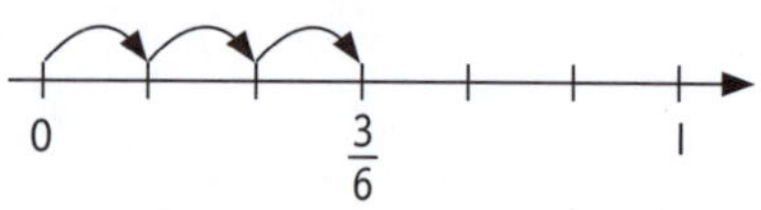

6

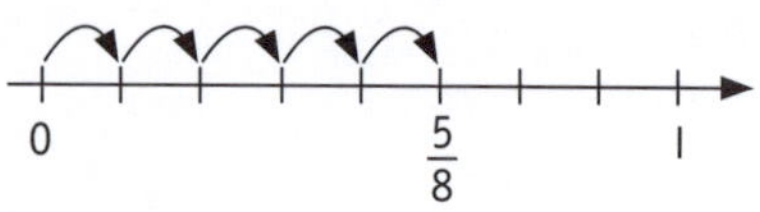

Quick Check

1 $\frac{1}{3}$ 2 $\frac{4}{5}$ 3 $\frac{7}{8}$ 4 $\frac{3}{4}$ 5 $\frac{4}{6}$ 6 $\frac{7}{10}$

Adding Strings of Numbers

1 30 + 40 + 20 = 90; 3 + 4 + 1 = 8 → 98
2 40 + 20 + 10 = 70; 7 + 2 + 3 = 12 → 82
3 20 + 20 + 30 = 70; 5 + 7 + 3 = 15 → 85
4 20 + 40 + 20 = 80; 8 + 3 + 6 = 17 → 97
5 50 + 30 + 20 = 100; 2 + 4 + 6 = 12 → 112
6 70 + 30 + 40 = 140; 2 + 1 + 7 = 10 → 150

UNIT 28 STATISTICS & PROBABILITY

The Traffic Survey

1 7:00–8:00
2 8:00–9:00
3 Because many people go to school and work at that time
4 Cars
5 115

UNIT 28 MEASUREMENT & SPACE

Stacking Blocks

1 3 2 4 3 5 4 6 5 16 6 16

UNIT 28 NAPLAN*-STYLE QUESTIONS

1 $\frac{4}{6}$ 2 80 3 11

UNIT 28 PROBLEM OF THE WEEK

Work Smart, not Hard

Luis's Strategy

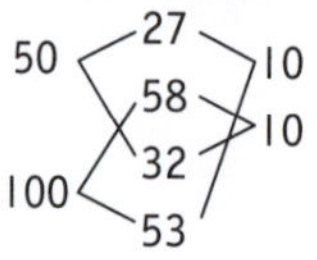

150 + 20 = 170

Max's Strategy

27 + 53 = 80
58 + 32 = 90
→ 170

TARGETING MATHS HOMEWORK: YEAR 3 © PASCAL PRESS ISBN: 9781925726565

Kiku's Strategy

27 + 58 + 32 + 53

20	+	50	+	30	+	50	=	150		170
7	+	8	+	2	+	3	=	20		

UNIT 29 NUMBER & ALGEBRA

Multiplication Turn-around Facts

1 12, 12
2 35, 35
3 40, 40
4 32, 32
5 18, 18
6 24, 24
7 24, 24
8 27, 27
9 24, 24
10 36, 36
11 28, 28
12 55, 55
13 56, 56
14 72, 72

Quick Check

1 12
2 25
3 18
4 50
5 35
6 16
7 28
8 40
9 18
10 60
11 70
12 10
13 36
14 45
15 100
16 16
17 21
18 40
19 90
20 16

Multiplication and Division Fact Families

1 28 ÷ 4 = 7
2 28 ÷ 7 = 4
3 4 × 7 = 28
4 7 × 4 = 28
5 18
6 3
7 18
8 6

UNIT 29 STATISTICS & PROBABILITY

A Chance Experiment

Open-ended, but expected values would be:
A 12 times, B 8 times and C 4 times.

UNIT 29 MEASUREMENT & SPACE

Fractions of a Measurement

1 $1\frac{1}{2}$ kg, 1500 g
2 $\frac{3}{4}$ L, 750 mL
3 $1\frac{1}{2}$ m, 150 cm
4 $1\frac{1}{4}$ L, 1250 mL
5 $2\frac{1}{2}$ kg, 2500 g
6 $1\frac{1}{4}$ m, 125 cm
7 1250 g

UNIT 29 NAPLAN*-STYLE QUESTIONS

1 5
2 Third option
3 $24

UNIT 29 PROBLEM OF THE WEEK

Mamma Bella's Cafe

Bella's Mains List
Pasta salads: 18
Burgers: 15
Pizzas: 16

Bella's Dessert List
Ice-cream sundaes: 15
Fruit salads: 18
Apple crumbles: 16

Mamma Bella's Cafe Takings
49 @ $10 = $490
49 @ $5 = $245
Total: $735

UNIT 30 NUMBER & ALGEBRA

Rules for Multiplication

1 2 × 4 + 4 = 12
2 2 × 5 + 5 = 15
3 2 × 6 + 6 = 18
4 2 × 7 + 7 = 21
5 2 × 8 + 8 = 24
6 2 × 9 + 9 = 27
7 2 × 6 + 2 × 6 = 24
8 2 × 7 + 2 × 7 = 28
9 2 × 8 + 2 × 8 = 32
10 2 × 9 + 2 × 9 = 36
11 5 × 3 + 1 × 3 = 18
12 5 × 7 + 1 × 7 = 42
13 5 × 8 + 1 × 8 = 48
14 5 × 9 + 1 × 9 = 54

Quick Check

1 4
2 15
3 36
4 50
5 12
6 24
7 27
8 18
9 70
10 18
11 24
12 18
13 33
14 44

Addition and Subtraction

1 53 + 30 + 6 = 89
2 44 + 30 + 3 = 77
3 54 + 20 + 5 = 79
4 38 + 20 + 3 = 61
5 76 − 20 − 4 = 52
6 49 − 30 − 1 = 18
7 83 − 50 − 1 = 32
8 108 − 80 − 2 = 26

UNIT 30 STATISTICS & PROBABILITY

Probability

Open-ended

UNIT 30 MEASUREMENT & SPACE

The Hands on the Clock

1 Close to a right angle

3 Close to a right angle

2 Less than a right angle

4 Larger than a right angle

5 Less than a right angle

UNIT 30 NAPLAN*-STYLE QUESTIONS

1 $108
2 $42
3 Quarter to six

UNIT 30 PROBLEM OF THE WEEK

Problem of the Week

Ada: 42 + 63 + 15
Bindi: 14 + 25 + 36
Luis: 61 + 52 + 43
Max: 61 + 42 + 35
In each case, other possibilities can be found by changing the position of the digits in either or both the 10s and 1s positions. For example, Ada's total is the same with 43 + 65 + 12 or 65 + 34 + 21.

UNIT 31 NUMBER & ALGEBRA

Small and Large

1 Smallest: 7658.
Largest: sixty-three thousand, five hundred and four
2 Smallest: 30 946.
Largest: thirty-six thousand and forty-nine
3 Smallest: 78 045.
Largest: seventy-eight thousand, five hundred and four
4 Smallest: 30 946.
Largest: thirty-six thousand and forty-nine
5 Smallest: 30 956.
Largest: thirty-six thousand, eight hundred and fifty-three
6 Smallest: 178 504.
Largest: one hundred and eighty thousand, nine hundred and thirty-seven
7 Smallest: 205 736.
Largest: two hundred and six thousand, one hundred and two
8 36
9 6
10 123
11 10
12 1
13 0

Quick Check

1 10 000
2 13 790
3 100 000
4 10 009
5 54 808
6 100 009
7 13 956
8 25 056
9 100 099
10 28 659
11 30 656
12 100 999

Change from $100

1

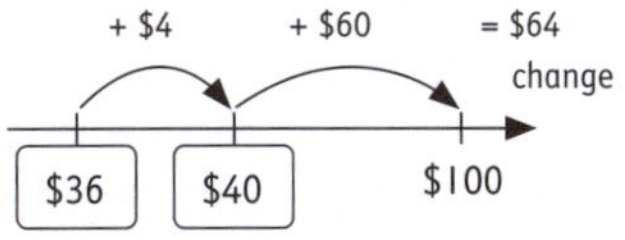

2

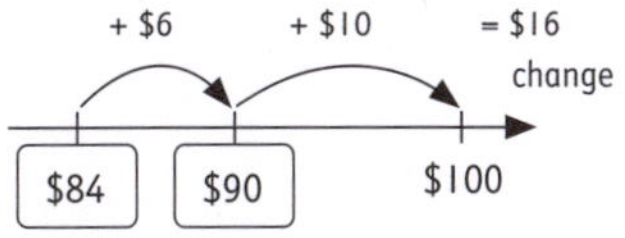

3
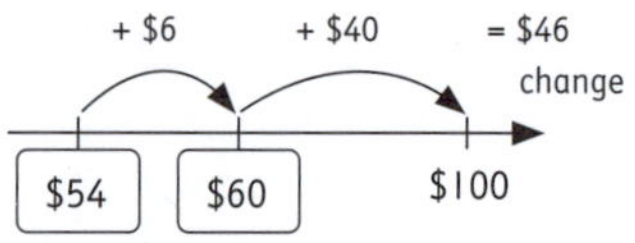

4
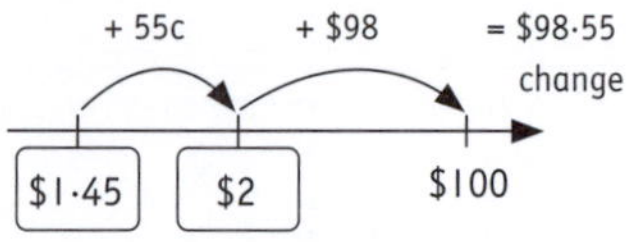

5
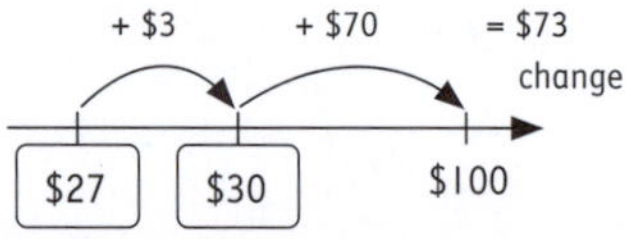

UNIT 31 STATISTICS & PROBABILITY

The Lion Cubs

1 Eunice
2 Vera
3 Eunice and Wallace
4 Eunice and Wallace
5 2900 g

UNIT 31 MEASUREMENT & SPACE

Time Spans

Approximate answers:

1 2 min
2 45 min
3 2 h
4 $1\frac{1}{2}$ min or 90 sec
5 30
6 12 h
7 30 days
8 365
9 12
10 90

UNIT 31 NAPLAN*-STYLE QUESTIONS

1 $2·80
2 865 420
3 There are 6 black cats.

UNIT 31 PROBLEM OF THE WEEK

Fun with a Calculator

1 These moves can be made in any order.
Move 1: – 700 000 leaves 36 859
Move 2: – 30 000 leaves 6 859
Move 3: – 6000 leaves 859
Move 4: – 800 leaves 59
Move 5: – 50 leaves 9
Move 6: – 9 leaves 0
2 Many possibilities – check moves on a calculator

UNIT 32 NUMBER & ALGEBRA

Estimation for Division

1 6 × 20 = 120, more than 20, 132 ÷ 6 = 22
2 3 × 30 = 90, less than 30, 84 ÷ 3 = 28
3 4 × 20 = 80, less than 20, 76 ÷ 4 = 19
4 5 × 20 = 100, less than 20, 95 ÷ 5 = 19
5 4 × 30 = 120, more than 30, 132 ÷ 4 = 33
6 6 × 10 = 60, more than 10, 78 ÷ 6 = 13
7 7 × 20 = 140, more than 20, 154 ÷ 7 = 22
8 9 × 10 = 90, more than 10, 108 ÷ 9 = 12
9 3 × 100 = 300, less than 100, 294 ÷ 3 = 98
10 6 × 20 = 120, more than 20, 138 ÷ 6 = 23

Quick Check

1 24 ÷ 6 = 4, 4 × 6 = 24
2 21 ÷ 7 = 3, 3 × 7 = 21
3 32 ÷ 4 = 8, 8 × 4 = 32
4 54 ÷ 9 = 6, 9 × 6 = 54
5 35 ÷ 5 = 7, 5 × 7 = 35
6 48 ÷ 4 = 12, 12 × 4 = 48
7 27 ÷ 9 = 3, 3 × 9 = 27
8 36 ÷ 4 = 9, 4 × 9 = 36
9 45 ÷ 9 = 5, 5 × 9 = 45
10 44 ÷ 4 = 11, 11 × 4 = 44

Estimation for Multiplication

1 Estimate: 80 × 3 = 240
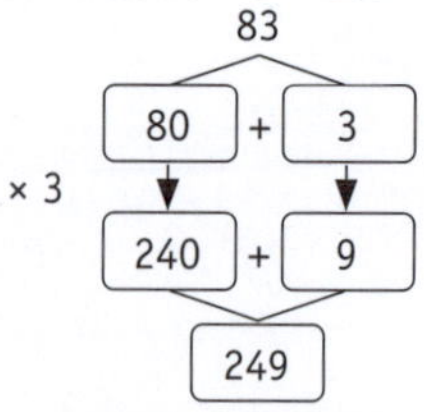

2 Estimate: 70 × 4 = 280
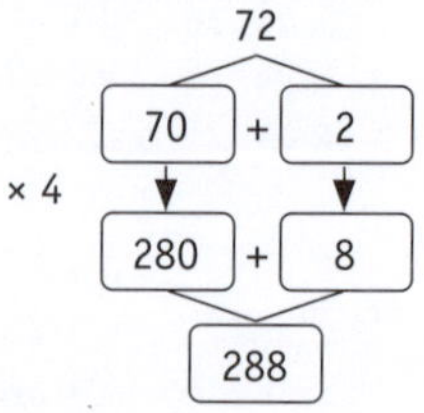

3 Estimate: 40 × 4 = 160
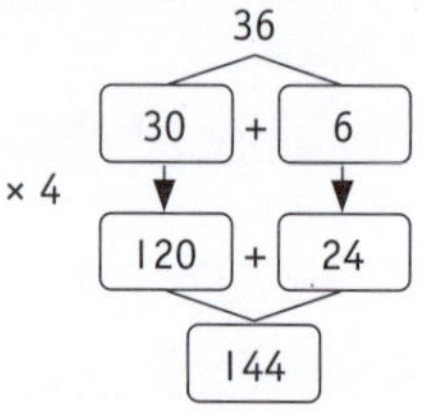

4 Estimate: 20 × 6 = 120
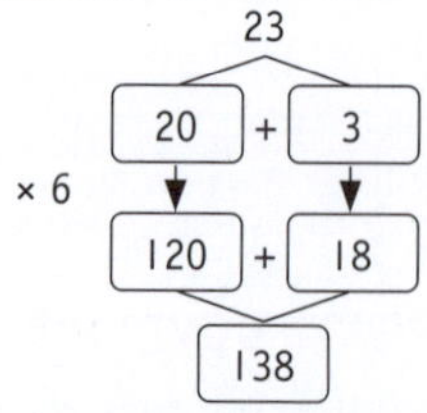

5 Estimate: 40 × 3 = 120
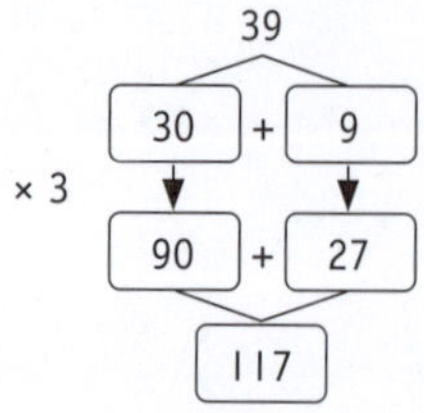

6 Estimate: 110 × 4 = 440
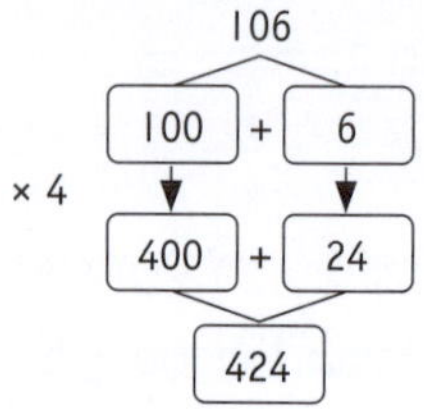

UNIT 32 STATISTICS & PROBABILITY

Frequently Used Letters

A = 23, O = 7, E = 21
1 Max
2 More tallies
3 U has a similar frequency to O and the frequency of I is slightly higher than O.

UNIT 32 MEASUREMENT & SPACE

Giving Change

1
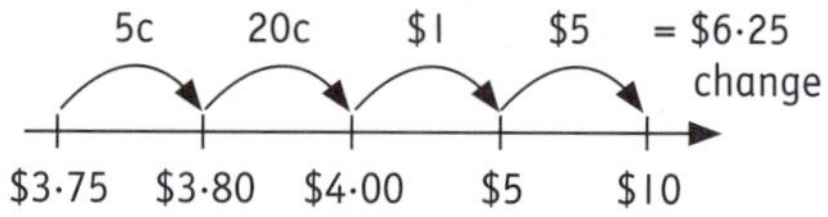

2
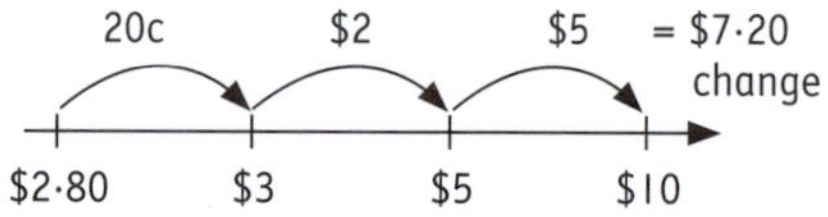

3
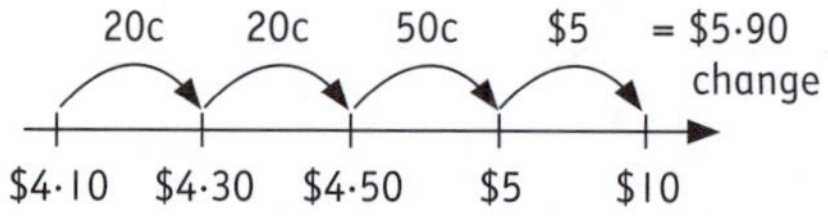

4
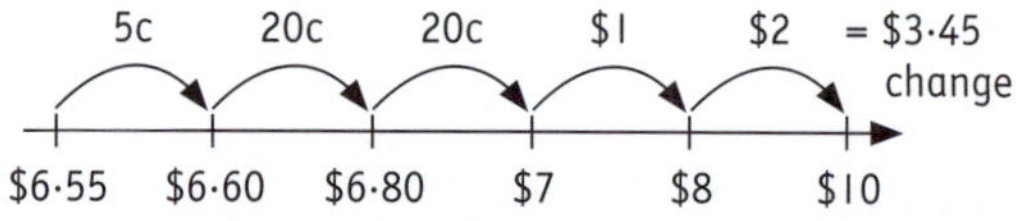

5
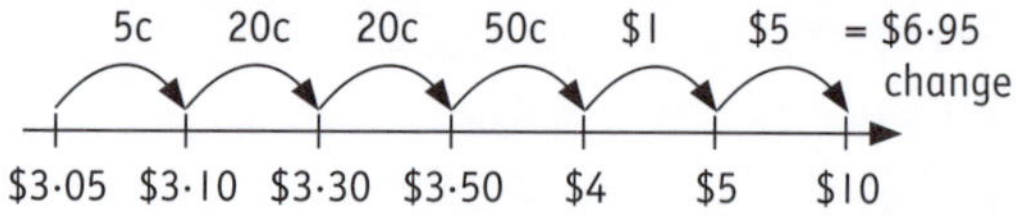

UNIT 32 NAPLAN*-STYLE QUESTIONS

1 6
2 6 × 8 = 48,
8 × 6 = 48
3 $2·80

ANSWERS

TARGETING MATHS HOMEWORK: YEAR 3 © PASCAL PRESS ISBN: 9781925726565

UNIT 32 PROBLEM OF THE WEEK

Multiplying Dice Game

This table show which products can be made.

×	1	2	3	4	5	6
1	1	2	3	4	5	6
2	2	4	6	8	10	12
3	3	6	9	12	15	18
4	4	8	12	16	20	24
5	5	10	15	20	25	30
6	6	12	18	24	30	36

18 of the numbers 1–36 can be made as the product of two dice numbers and 18 numbers cannot be made.

TERM 4 REVIEW

Review: Unit 25 and Unit 26

1 25 672
2 50 465
3 18 076
4 38
5 20
6 35, 37, 39
7 62, 60, 65
8 20 min + 15 min = 35 min
9 15 min + 30 min = 45 min
10 $\frac{3}{4}$
11 $\frac{5}{8}$
12 11 stickers
13 Luis: 11 strawberries
Bindi: 17 strawberries
14 **Coloured Counters**

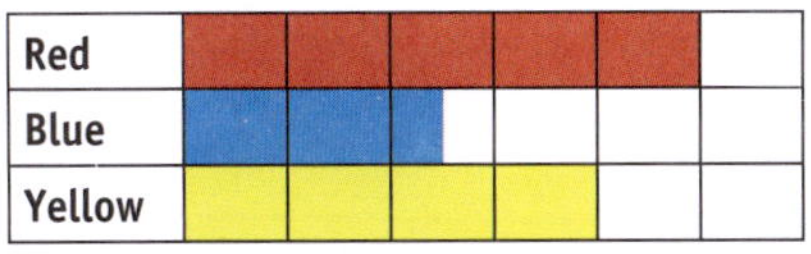

Review: Unit 27 and Unit 28

1 Estimate 20, Actual 18
2 760
3 700
4 900
5 1000
6 1400
7 5 × 8 = 40
8 100 − 64 = 36
9

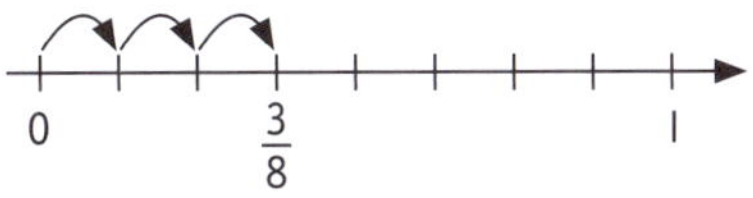

10

11 40 + 20 + 30 = 90; 3 + 5 + 1 = 9 → 99

12 30 + 40 + 30 = 100; 3 + 4 + 2 = 9 → 109

13 4
14 12
15 6
16 12

Review: Unit 29 and Unit 30

1 3 × 6 = 18, 6 × 3 = 18
2 7 × 4 = 28, 4 × 7 = 28
3 4 × 9 = 36, 9 × 4 = 36
4 3 × 7 = 21, 7 × 3 = 21
5 4 × 9 = 36 onions
6 24 ÷ 6 = 4 slices
7 1500 g
8 150 cm
9 2500 mL
10 3 × 6 = 2 × 6 + 6 = 18
11 3 × 9 = 2 × 9 + 9 = 27
12 4 × 7 = 2 × 7 + 2 × 7 = 28
13 4 × 9 = 2 × 9 + 2 × 9 = 36
14 44 + 50 + 3 = 97
15 62 + 40 + 8 = 110
16 87 − 50 − 3 = 34
17 96 − 50 − 7 = 39
18 4:10, No
19 9:00, Yes

Review: Unit 31 and Unit 32

1 65 472
2 13 064
3 7759
4 13 858
5 18 056
6 \$27 + \$3 + \$70 = \$100
so \$73 change
7 \$43 + \$7 + \$50 = \$100
so \$57 change
8 500 g
9 700 g
10 300 g
11 25
12 15
13 12
14 27 ÷ 3 = 9, 3 × 9 = 27
15 36 ÷ 4 = 9, 4 × 9 = 36
16 Estimate: 50 × 6 = 300

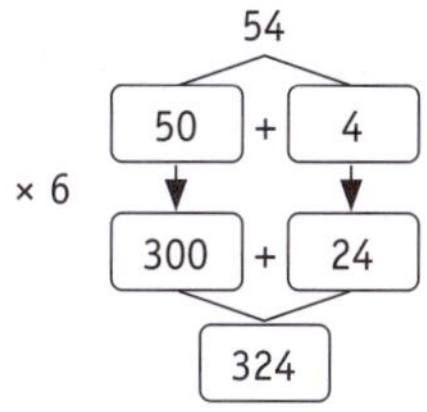

17 Estimate: 40 × 5 = 200

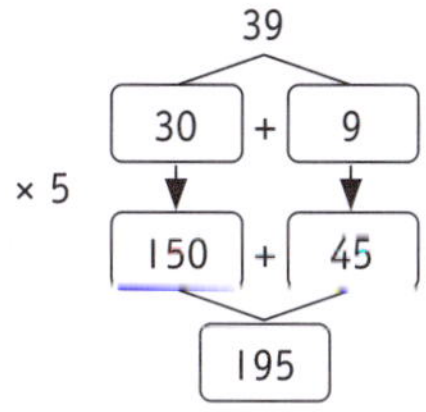

ANSWERS